Jade Fever

Hunting the Stone of Heaven

Stan Leaming
with Rick Hudson

Heritage
House

Heritage House Publishing Co. Ltd.
#108-17665 66A Ave.
Surrey, BC, Canada   V3S 2A7
greatbooks@heritagehouse.ca
www.heritagehouse.ca

Library and Archives Canada Cataloguing in Publication
Leaming, Stan
    Jade fever: hunting the stone of heaven / Stan Leaming with Rick Hudson.

Includes bibliographical references and index.
ISBN 1-894384-85-7

    1. Jade. 2. Leaming, Stan. I. Hudson, Richard (Richard Dennis), 1947- II. Title.

QE394.J3L42 2005   553.8'76   C2005-900487-8

Front-cover photo: Stan Leaming
Back cover: *Silent Reflection* by Lyle Sopel, courtesy of Lyle Sopel Studio/Gary Wildman
    Photography
Cover and book design/layout by Nancy St.Gelais
Edited by Catherine Johnson
This book is set in Adobe Garamond.
Printed in Canada

Heritage House acknowledges the financial support for its publishing program from the Government of Canada through the Book Publishing Industry Development Program (BPIDP), The Canada Council for the Arts, and the Province of British Columbia through the British Columbia Arts Council.

The Canada Council | Le Conseil des Arts
    for the Arts | du Canada

BRITISH COLUMBIA
ARTS COUNCIL
Supported by the Province of British Columbia

To all those rockhounds and jade enthusiasts who contributed to my store of information on minerals: thank you.

Stan Leaming

Table of Contents

Preface

Retired now for more than 20 years, I feel the need to record something of my professional life, particularly my time from 1960 to 1982 with the Geological Survey of Canada (GSC). Before joining, I spent 12 years in mineral exploration, from Labrador in the east to the Yukon in the northwest, and in all that time and over all that area I had not been aware of the mineral jade. However, once I'd signed on with the Survey, this changed rapidly. British Columbia was then at the apex of the lapidary hobby boom, popularly called rockhounding. Rockhounds prized agate, petrified wood, and pre-eminently jade as materials from which to craft jewellery. It turned out that jade boulders were relatively common along the banks of the Fraser River, from Lillooet to Hope in southern British Columbia, and during my time at the GSC, the source of these boulders was discovered in the upstream mountains. Once the conditions of provenance were known, many other sources were discovered.

I was trained as a geologist and was given, accordingly, the opportunity to study the geological provenance of all of these deposits. The result was a publication entitled, rather grandly, *Jade in Canada*.[1] This gave me some international publicity, as there was little else in print on the subject. In due course, it came to the attention of Dr. Chris Nagle of the Smithsonian Institution, who invited me to search for the source of Inuit jade artifacts found in Labrador. Later, Mr. Russell Beck, curator of the Southland Museum and Art Gallery in Invercargill, New Zealand, saw the publication and thought I would be interested in a trip to China to see the jade deposits in the far western province of Xinjiang. While on that tour we also visited jade deposits in Taiwan and South Korea. Some years after that, Russell organized a trip to Siberia to look at deposits in the East Sayan Mountains. Then, in 1999, I satisfied a long-term wish to visit the jade deposits in South Australia and New Zealand.

At the outset I decided that *Jade Fever* would not be strictly technical, as was *Jade in Canada*, but would be more personal and include some of my memories pertaining to two decades with the GSC. And time is growing short. I will be 100 years old in little more than a decade, and I have noticed that at the end of one's life, time speeds up at the same rate as one's mental processes slow down. Perhaps this is Leaming's Law.

All that remains is to thank those pioneers who made possible the jade story as I tell it. Many, alas, are no longer with us. But they are included here, with my sincere appreciation: Joe Bell, Bud Davidson, Gerry Davis, Robert Dubé, Walter Ellert, Clancy Hubbell, Don and Gwen Lee, Jeannie MacCulloch. Ruth McLeod, Oscar Messerer, Ed Osterlund, Frank Plut, Stan Porayko, Alex Schick, Ben Seywerd, Bob Smith, Lyle Sopel, Cleo and Ned Sparkes, Bill Storie, Harry and Nellie Street, Ed Tucker, Deborah Wilson, David and Kwai Wong, Bill Yarmack, Guy Yarmack, and Bob Yorke-Hardy.

Special thanks are due to Marion Scott, whose interest in marketing jade sculptures helped many B.C. artists. Also, thanks to David Saxby and Gary Gallelli, whose acumen predicted the potential of the whole jade business. I am grateful to Dr. R.G. Coleman of the U.S. Geological Survey, who gave me much valuable technical advice, and to Winnie Robertson and Kirk Makepeace, both of whom assisted with written reminiscences and other memorabilia that jogged a fading memory, and for checking the text. I am particularly indebted to Russell and Ann Beck, who gave invaluable advice on the New Zealand section. Thanks also to Tom Vaulkhard for proofing the manuscript on its completion. Finally, many thanks to my editors, Catherine J. Johnson, Karla Decker, and Vivian Sinclair, for taking the manuscript and producing a most polished product from something that started out as a diamond in the rough.

Stan Leaming
Summerland, B.C., August 2004

Introduction

In 2001 friends of mine moved to Summerland in British Columbia, and in due course I received an invitation to visit their new home. While the request was welcome, there was another reason for my eagerness to stay over in that pretty little town on Okanagan Lake: I knew that Stan Leaming lived there.

Some years earlier I had managed, through dint of perseverance, to acquire in second-hand bookshops both his *Rock and Mineral Hunting in British Columbia* and his *Jade in Canada*. Both, alas, are long out of print but are considered classics by anyone interested in exploring the rich mineral diversity of the province. It was with some uncertainty that I called up Stan out of the blue and asked if he would autograph my two copies. This he graciously did, but our meeting turned into much more than a master-and-student encounter. It was the start of a friendship that began with a shared passion for geology but quickly moved on to something even better—a common project. During that initial visit, Stan took me to his office, a semi-chaotic room packed with the memories of half a century of geologizing. Books, reports, slides, equipment, samples, and computer peripherals covered every level space. Slabs of rock, photographs, magazines and, yes, jade were stacked everywhere. Almost immediately Stan posed the question, "What am I going to do with all this stuff?" What indeed.

Stanley Fraser Leaming was born in Minnedosa, Manitoba, in 1917, where he attended school. He later moved to Brandon with his family and graduated in 1936. The Depression still held the region in its grip, and there were no jobs for lads. Stan, a frail youngster interested in electricity and geology while most of the other boys played football and hockey, was encouraged by his CPR-employed father to stay out of the job hunt. So Stan's parents paid for him to complete Grade 12, as it was then considered the first year of university.

Although geology intrigued Stan, he had little idea of what it entailed. As a boy on holiday at his grandfather's home in Kenora, Ontario, he remembered prospectors showing him samples. It was likely the excitement from those days, plus the thought of an outdoor life, that tipped the scales in favour of geology.

The government of Manitoba had started a prospecting school at Falcon Lake under Dr. G.M. Brownell, and Stan was accepted. The students there learned to prospect, drill, and blast. The program terminated with the declaration of war in September 1939. However, Stan's father insisted that he finish his B.A., which he did at Brandon College under Dr. John Evans.

To a young man with prospects but no job, the war seemed an exciting chance to get away, and on November 18, 1940, shortly after his 23rd birthday, he signed on with the Royal Canadian Air Force (RCAF). Being technically inclined, Stan trained in radio communications and, later, on the new (and very secret) RDF, an early form of radar. The

At prospecting school near Brandon in 1939, Stan (second from left) learned basic geology, prospecting, and blasting.

year 1941 saw him in England, Wales, and Scotland, tracking incoming German bombers. He was a participant in the struggle for control of the skies over Britain.

In 1942 Stan shipped on a convoy to Malta. It was en route that he probably came closest to "catching it," when a German JU-88 bomber swooped in on the stern of his ship. From his 12-pounder gun position, he recalls watching the plane almost with disinterest, seeing the bomb separate from the aircraft and glimpsing the pilot's goggled face as he banked away. Then Stan heard the huge splash as the bomb fell just short of the ship's stern and spray flew high into the air, soaking him. "I didn't feel fear," he said later. "At that age, you think you're immortal, and it can't happen to you."

When the Allies swept up Italy, Stan expanded his European experience with sightseeing trips to Naples and then to Rome. He later saw the southern France landings from a radar ship. On August 31, 1945, Sergeant Leaming was discharged from the RCAF and returned to Canada, along with tens of thousands of other personnel. With a serviceman's grant he enrolled at the University of Toronto, whence he emerged in 1948 with an M.A. in Geology.

Thus began the life of a field geologist. There was plenty of work, and Stan travelled the length of Canada, prospecting and consulting. Those were boom times, as the country shook off the gloom of war and industry grew. While prospecting in B.C. he met Kay Spall, a teacher at Clearwater School, near where he was stationed. A year later, on December 31, 1955, they married in Kay's hometown of Kelowna, B.C. Stan was 38, Kay "a lot younger." The following five years saw Stan and Kay travelling all over the country as geology boomed and busted. The family grew with the arrivals of Ruth, Christopher, Charlotte, and, finally, Catherine.

In 1960 Stan signed on with the Geological Survey in Vancouver under Dr. Jack Armstrong. His role was information officer, notably for the growing ranks of rockhounds throughout the province. A meticulous note-taker, he began to document the resources in the province that were of interest to amateur collectors, and in 1971 the GCS published his *Rock and Mineral Collecting in British Columbia*, which is still today the most detailed book on rockhounding in the province.

The jade industry was beginning to boom as first Taiwanese and then Chinese buyers came looking for the "stone of heaven." Stan spent time with the newly established mines, first in the Lillooet area, then in the Omineca, where O'Ne-ell Creek and Ogden Mountain became well known as one jade discovery followed another. From 1970 to 1975 he was based near Dease Lake, where he mapped the ultramafic belts of northern B.C., covering the newly discovered jade deposits of Cassiar, Provencher Lake, Kutcho Creek, and Letain Lake.

Further work led to his publishing *Jade in Canada* in 1978, another definitive reference book. While miners like Larry Bell, Howard Lo, and Don Lee were putting B.C. jade on the world map commercially, Stan Leaming's tightly written book gave the industry a technical depth that buttressed its position on the world stage and turned B.C. into the jade capital of the world. In 1980, with son Chris as photographer, he produced *Guide to Rocks & Minerals of the Northwest*. Such was its success that it is still in print over two decades later.

In 1981 Stan retired from the Survey, and he and Kay moved to Summerland, B.C. But in retirement Stan was only just getting started. Consulting work followed, and later he teamed up with a number of international groups of jade experts to visit Canada's Labrador coast, the legendary Mountains of Kunlun in western China, Korea, Taiwan, Japan, and Lake Baikal in eastern Russia. There were days of bad food, bizarre customs, and barbaric travel, but discomfort never fazed Stan, and his diaries of those trips are filled with minutiae describing everything from an orange-juice box on a plane to the number of children in an Uighur village.

As I looked into Stan's office in Summerland in 2001, I saw the memories of someone who had worked as the provincial jade geologist for over two decades, during that critical period when the jade sector began, grew, imploded, and finally developed into the world-leading business that it is today. In one room were the reports, the stories, the details that turn history into folklore.

There was no way that I, as a rockhound and writer, could walk out of that office without making a commitment. This book is the result. Stan Leaming stands alone in the province for his contribution to helping the prospectors and other "little guys" with his fieldwork, his reports, and his two books that put B.C. mineral collecting on the map. He was the right person, in the right job, at the right time.

Rick Hudson
North Saanich, B.C.
August 2004

The Canadian Jade Story

 The Chinese called it *yu*, the Maoris called it *pounamu*, the Aztecs called it *chalchiutl*, and the Spanish called it *yjada*. We call it jade.

So what is jade? Simply put, it's an aggregate, or a blend of minerals. The word "jade" is actually a generic term used to describe two distinct mineral aggregates: nephrite and jadeite. However, since "nephrite" and "jadeite" have been synonymous in the English language since the words arrived in the early seventeenth century, they will be used interchangeably throughout the book, although this usage is not strictly correct except when talking about jade in British Columbia, where all "jade" is indeed nephrite. (For an in-depth explanation, see "The Science of Jade.")

B.C. is the jade province par excellence. In fact, if you talk about Canadian jade you could almost be talking about B.C. jade; nowhere else in Canada is there such a concentration of the mineral. Jade has also been found in the Yukon, the Northwest Territories, and Newfoundland— it can form in any region with the requisite geology. Because of jade's relatively minor importance as a mineral, records on it have only been kept from 1963, although statistics on gold, copper, zinc, etc. go back over a century.

After the conclusion of the Second World War, there emerged in the United States and Canada (as well as elsewhere in the world) a fraternity of amateur lapidaries who, for some quirky reason, called themselves "rockhounds." While there were a few loners, rockhounds for the most part, then as now, were a gregarious, friendly group devoted to collecting minerals and finding, cutting and polishing rocks that had some attractiveness of colour or pattern. The polished specimen could be used for anything from bookends to ring stones. Many rockhounds became accomplished jewellers, mounting their handicrafts in silver or even gold settings. Others were interested in the mineral kingdom and became knowledgeable in the many varieties and specimens of minerals. Others collected fossils. Part of the appeal was that there was something for everyone.

Active in clubs, rockhounds shared their knowledge of localities where minerals could be collected and, with clubhouse facilities, lessons in polishing rocks were open to all members. The use of communal equipment meant that no one needed to make any great investment to participate in the hobby. At one time there were 40 clubs in British Columbia. Ontario and Alberta also boasted many societies, and the other provinces with smaller populations were represented in proportion.

The hobbyists supported a magazine for nearly a quarter of a century: the *Canadian Rockhound* was founded in 1957 and flourished under the guidance of Cleo Sparkes. The principal contributors were mostly from British Columbia—to such an extent that it might well have been called the *B.C. Rockhound*. The first issue cost 35 cents. When the magazine came to an end (at which time it cost 65 cents), it was replaced for a while by *Cab and Crystal*, which was mainly the work of Ms. M.L. Fraser and ran from March 1988 to the spring of 1992. In 1998, aware that much of the rockhound "action" occurred on the west coast, Ms. Win Robertson began editing the *B.C. Rockhounder*, which continues at this time.

~ British Columbia ~

There is a wealth of lapidary-quality material in British Columbia. Agate, jasper, rhodonite, and jade are relatively abundant. Carnelian, amethyst, nickel silicate, smoky quartz, and garnets can be found too. Along the Fraser River, gravel bars that witnessed the start of the gold rush back in 1858 are once again occupied by prospectors, heads bowed down, scanning the pebble piles for suitable material. In 1968 Premier W.A.C. Bennett decreed that jade was the official provincial stone, and that anyone could collect it along the Fraser, provided they did so for their own recreation only. This mineral reserve remains to this day.

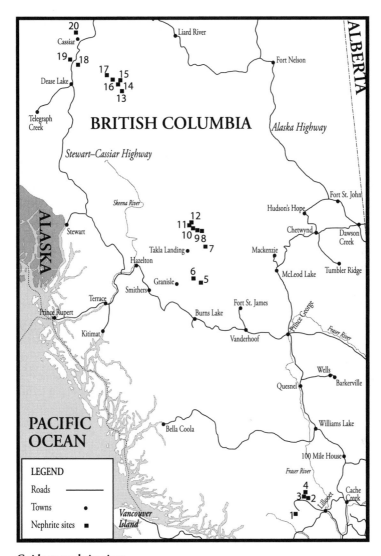

Guide to nephrite sites:
1 Noel Creek/Royal Jade. 2 Hell Creek/Birkenhead. 3 Jim Creek. 4 Blue Creek. 5 Mt. Sidney Williams.
6 Jade Queen/O'Ne-ell Creek. 7 Kwanika Creek. 8 Vital. 9 Quartz Creek. 10 Fran 3. 11 Mount Ogden.
12 Ogden Creek. 13 Kutcho Creek/Cry Lake. 14 Provencher Lake. 15 Letain Creek. 16 Wheaton Creek.
17 Polar Jade/Serpentine Lake. 18 Greengold. 19 Seywerd Creek. 20 Cassiar Mines.

Map 1: *The principal nephrite jade belts in British Columbia are concentrated around three zones. In the south is the Lillooet District, in the centre the Omineca District, and in the north the Dease Lake and Cassiar region.*

The Lillooet jade pioneers

A 1961 account by Stuart Holland[1] for the B.C. Ministry of Mines brought to the attention of many potential jade prospectors the key information for hunting this exotic mineral. In his report Dr. Holland mentioned the association of jade with serpentine, the latter being a relatively common metamorphic rock. The basic premise was simple: everywhere serpentine was found, prospectors should search for jade. Map 1 shows the distribution of the jade fields, with respect to serpentine rocks. Within British Columbia, it transpired that there were three main regions: the Lillooet area, the Omineca area, and the Cassiar region. Lillooet was historically the first jade zone to be explored, but it was not long before the central and northern belts came into their own, and today the main jade production is from those areas.

Probably the real pioneers of the Lillooet area were two people mentioned by Holland, C.J. Hallesay and Captain Duncan, who collected along the river before 1950. Their discoveries attracted many other locals, and the river became known as a good source of alluvial (river) boulders of reasonable size. However, many jade finds were simply too big for easy collection. Some along the Bridge River weighed up to 20 tons.

Rockhounds have contributed much since early on. One of the players during those initial years was Wilf Zacharius, who uncovered alluvial jade deposits along the Bridge River in the early 1960s. Wilf was a dealer as well as a rockhound, and many of the boulders along the Bridge yielded to his saw. He went into business with his son, calling the business Zac 'n Zac. He once recalled buying a boulder, untested, for $500—a lot of money then—but there was not a pound of good material in the whole thing. After that, he would only purchase material once a test cut was made with a portable saw that he designed himself.

Huge boulders were not a problem for Bill Yarmack,[2] initially a Vancouver plasterer by trade who worked with his stepson Bud Davidson. After buying his first boulder (which weighed over a ton) from someone near Lytton, he cut it with his own 30-inch saw. Later he headed down into the States, trading blocks of jade for any rock material he could find. When he reached El Paso in Texas, he turned around and went home and, with his accumulated trades, opened a rock shop. The saw that Bill developed would probably have been one of the earliest models. It was a drop saw on a boom, allowing for easier alignment and cutting of large shapes. He also used water as the cooling agent, rather than the more expensive and messy cutting oil.

Bill later began marketing jade through his rock shop in Chase, B.C., where he moved in 1957. He successfully sold jade to the Germans, where the famous Idar–Oberstein factories had been turning out lapidary products for many centuries. He also sold to the Orient and was one of the first to realize that to make a living in the jade business he needed to control sales and marketing. Like Wilf Zacharius, Bill learned a few tough lessons when he started, and not all his finds were successful. He once paid $1,500 for a three-ton boulder that he later described as "a turkey."

In the 1960s the price per pound varied from 25 cents for junk jade to $30 a pound for the best quality. The value of a new find was always in the balance until the stone was cut. In reality the average price was seldom over "a buck a pound." A classic example of the "jade uncertainty principle" was when a mining syndicate found a 1.5-ton boulder in the remote Omineca Mountains in northern B.C. They skidded it out one winter, a hundred kilometres along frozen riverbeds, and then put it on a flatbed to Vancouver. During a glittering cocktail party at the ritzy Bayshore Inn it was presented to the press, with "experts" suggesting it was worth up to $30,000 (a king's ransom at that time). In due course, people asked Yarmack what he thought it was worth. He suggested two dollars a kilogram or $3,000 for the lot. "They wanted to throw me out," he recalled after. But two months later, a member of the syndicate called him to ask if that offer was still good! They couldn't find another buyer with a better price. "It was a nice boulder," Bill said. "I sold some of it for $25 a pound ($55 a kilogram), but it could just as easily have been a junker."

In those early days, expenses could be significant for jade hunters on the Fraser, Bridge, and Yalakom rivers. To pull a boulder out of the water required a wrecker (at eight dollars an hour), plus one dollar a mile to the site. The wrecker might need 2,000 feet of cable, and if the rock was big, a crane was necessary to lift it when it was at the road, and a flatbed was needed to haul it away. To cut it, the only saws were located at monument works, which might be in Vancouver—over 300 kilometres away. Costs mounted rapidly.

At one stage, Bill Yarmack was turning over $50,000 a year of jade—a tidy sum in the late '60s. Because the jade source had not yet been found in the Shulaps Mountains west of the Yalakom River, he had to rely on alluvial finds. Bill claimed the only thing that stopped him from doubling his business was the lack of raw material. He bought from numerous jade hunters, usually at one dollar a kilogram, but like all prospectors at heart, he dreamed of finding the motherlode himself. Ironically, within a year Ed Osterlund would do just that, but the increased supply simply depressed the market, and turnover remained much the same as before.

Harry and Nellie Street of Gold Bridge were avid prospectors. They became involved in 1965 with the discovery of a four-ton boulder of exceptionally high-quality jade,[3] which they managed to drag behind Harry's bulldozer for 10 miles to their cutting shed. Despite having 13 children, Nellie was as involved as Harry when prospecting. They ran a thriving jade business, which brought them an annual income of about $40,000, with some of the better jade selling at five dollars a pound.[4]

Working in the area around Carpenter Lake, the Streets staked several jade claims and were among the first to develop lode claims in 1969. This was where actual mining was necessary to liberate the imprisoned stone. (The alternative was to find loose "float" boulders that had been carried downstream.) Attacking bedrock meant using heavy machinery—notably bulldozers—to clear away the overburden. As it happened, Harry had operated one for the Department of Highways at an earlier stage in his life.

Harry and Nellie formed Royal Jade Mines and sold a 3,000-pound river jade boulder to a New York buyer for $36,000. Harry found jade in Marshall Creek and in the Bridge River, while Nellie is credited with finding the nephrite outcrop near D'Arcy on Carpenter Lake, across the Cadwallader Range. The deposit was directly under the B.C. Hydro power lines, which prevented any extensive mining operations, and the property remains unproven to this day. The main value of this discovery was that it authenticated the extension of the jade belt south across the range.

The Streets were more than prospectors, though. In addition to selling raw jade, they ran a production shop that turned out finished jewellery and other goods. However, given the size of their family and the cost of maintaining a D-7 Cat, they probably needed every dollar they earned. Nothing if not flexible, Nellie was later a short-order chef in Princeton, and Harry sold ultralight aircraft.

Lillooet was a favourite place of rockhounds and prospectors. One such local character was Ed Osterlund, who owned a ranch at the junction of the Yalakom and Bridge rivers. There were alluvial jade boulders in the riverbeds as well as placer gold at Horseshoe Bend. Ed held claims there, and one year during the quieter winter months, while moving boulders with his tractor to get at the gold-bearing sands, he uncovered jade cobbles and boulders. Then one spring, while checking on his cattle at the head of Hell Creek, he came across a nephrite outcrop. This was significant, as it was the first *in situ* nephrite discovery and, more importantly, the jade was associated with a much larger deposit of serpentine, thus confirming Stuart Holland's 1961 description of how the mineral was formed. After taking out a few loads, he sold the claims to Oscar Messerer, another of the jade pioneers.

Oscar was an airline mechanic and a machinist by trade. He could construct anything from metal, so the lapidary business was right up his alley. He operated under the business name of Birkenhead Mines and B.C. Gem Supply and fabricated most of his own equipment, including cabbing machines (specialized grinders) for his workshop. Now with his own supply of jade, he started to produce jewellery for export, notably to the U.S. But in 1968 the Americans were at the height of their anti-communist fervour. Everyone knew that jade came from China, and China was "red." The U.S. wanted no part of "red" jade.

As a government employee, I was considered neutral, unbiased, and, of course, honest, so I became the official certifying agent who confirmed that finished goods shipped to the States were indeed products of Canada, made from Canadian material. In 1975, when international tensions had eased somewhat, Oscar moved his operation to Hong Kong, where he had a factory with over 100 employees. It didn't last too long however. Despite the low cost of labour, the cultural differences resulted in misunderstandings and poor production. He returned to Canada, where he went back to his old trade.

Another avid rockhound and jade collector was Bob Bouvette. Bob ran a motel in Lillooet and spent much of his spare time prospecting, especially along the Fraser and Bridge rivers. Later he moved south of Lillooet and opened a motel there, where he continued to hunt jade along the rivers. After his wife died he moved to the Similkameen River east of Princeton to continue operating a rock shop. He brought along his jade stockpile, which often confused visitors who didn't know Bob's history in Lillooet. There is no jade to be found in the Similkameen.

Ron Purvis was a fruit farmer and lapidarian on a property beside the Fraser River south of Lillooet. He was a Second World War veteran and an army physical education expert who worked at St. George's Residential School for Native youth at Lytton. He often received jade boulders from the Native children in his classes. During his 14-year stay there he became interested in the rocks of the region—it is famous for its agates—and

On the flanks of the Shulaps Range above Hell Creek, Oscar Messerer loads blocks of jade prior to slinging them out in a net under a helicopter. The high cost of aircraft time, combined with the rugged terrain, limited jade production. Some of the best material was found loose in the talus slopes high on the mountain.

especially the jade boulders along the Fraser. It was Ron who found the famous "Buddha's Foot" boulder in 1968, which is now in the National Museum in Ottawa. In his book *T'shama*,[5] Ron refers to the deposit of green vesuvianite at Skihist Mountain just west of Lytton as "mountain jade," but it is unclear what he knew about the jade found along the riverbanks.

Then there was Bob Smith, an Ontario prospector, who hunted a tributary of the Bridge River known as Marshall Creek. Much of the area had already been staked, so he chose another tributary known as Brett Creek and, after finding a few cobbles that looked encouraging, hired a bulldozer and stripped the overburden for about half a mile up the riverbed. It's not clear whether he had a sixth sense, but he exposed a narrow vein of jade along a serpentine–chert contact.

Bob thought he had the world by the tail and promptly formed a company called Greenbay Exploration & Mining, hoping to set up a local factory to manufacture products rather than solely export the raw mineral. Several hundred tons of jade were mined, including jade from a neighbouring property run by his son. Eventually, though, the claims were sold to W. Schoenbaechler, who, with his wife Dolores, owned The Happy Prospector rock shop near Harrison Hot Springs in the 1990s. Some production followed this change of ownership, but the claims never became more than a passing interest. At one time Comaplex International Resources Ltd. took an interest in them, but they too did not make a success of the venture. Still, jade discoveries were popular with the news media, and a report in a Vancouver newspaper quoted Bob Smith as saying " ... they had traced the vein for three miles, and in the past month they had taken out about 200 tons of medium- and-gem quality jade."[6]

Other important players at that time were the husband-and-wife team of Don and Gwen Lee. They started their jade career by accident, at Bridge River, where Don was in charge of the machines on the new B.C. Hydro Carpenter Lake diversion. When the dam closed and the level of Bridge River dropped, Don noticed people in the creekbed, tapping rocks. He had no idea what they were doing but went and asked. That was how he learned about jade. His first Sunday out he found seven cobbles, one so large it had to be skidded out with a Cat. Things were easier in those days, and when the machines weren't in operation, there was a certain amount of leeway in allowing them to be used on other projects.

It turned out that Don had a good eye for the material, and one of those first discoveries was a high-quality nephrite boulder that got him hooked. Cutting and selling jade soon became much more interesting than operating a bulldozer, so he and Gwen opened a jewellery store in Langley—Lee's Jade & Opals—in 1965. In time, they came to an agreement with Bob Smith to market his jade worldwide, and they expanded into the German, U.S., and Taiwan markets. Don later became the sole agent for Mohawk Oil and between 1976 and 1981 routinely flew to the company's operation at Kutcho Creek, marking suitable pieces of jade for transport to their property in Langley while he was there. The product was then laid out for buyers, notably from Taiwan but increasingly from mainland China too.

Don and Gwen encouraged local sculptors to work with jade, and one of the artists in their stable was Robert Dubé, a true talent. They were proud that they could supply good-quality material to these artists in an effort to develop a local industry. A German businessman, Herr Wolf, approached the Lees to buy jade rough, but ended up buying four of their best jade carvings, including Robert Dubé's famous scalping sculpture. Gwen always maintained that Dubé was the greatest of all the jade artists. "He could carve anything," she said. The Lees once commissioned Robert to craft a gift for a dentist friend. It was to be a jade tooth set in gold, made with jade and gold that came from their own mining properties. When Robert completed the work and it was presented to the dentist, the tooth was so precise that the dentist identified it immediately.

In time, the Lees added gold mining to their activities and eventually ran a hydraulic operation in the Yukon, which Gwen later described in her book *Rivers of Gold: A True Yukon Story*. While Don and his son did much of the heavy work, Gwen was a remarkable businesswoman. Blessed with an easygoing nature, she successfully bridged the cultural divide that initially separated the Asian buyers from the North American jade

Helicopters provide a rapid although expensive way of getting jade to the roadhead and hence to the buyer.

producers. Every fall, some 200 to 300 tons of jade arrived at the Lees' yard in Langley, where Don graded it. Right from the start there were three main buyers from China who came every year. All of them, according to Don Lee, bargained hard, but a handshake was their bond, and they never changed the terms afterward. Initially a business relationship, it changed slowly over the years into a respectful friendship.

In the early '70s, the three men began talking to the Lees about buying into a jade mine. Don found that one of Ben Seywerd's properties had become available, so he registered an option. The Chinese trio gave the Lees $10,000 in cash to get the project started. Don got a rock drill and with his son, Brian, drove north to the property. Within a week Don knew the nephrite grade wasn't there. He returned $8,000 to the investors and never charged for his or Brian's time on the project. This honesty cemented a trust between buyer and agent that would last over three decades. In 2004 one of the buyers, Mr. T.C. Luo, was still visiting Vancouver every few months and never failed to either call the Lees or meet them for dinner.

Both the Taiwanese and Chinese buyers worked with cash. Gwen Lee recalls suitcases of the stuff placed on their kitchen table during negotiations. Over time, Gwen opened bank accounts for all three businessmen. She kept their books, signed their cheques and managed their accounts. Not one of them ever signed a document; such was their trust in Gwen's honesty.

Cecil McEwen became a jade prospector more or less by accident. He operated a tow truck for his garage business in Lillooet and was frequently hired to haul large boulders out of rivers for successful prospectors. In 1958, after a few such jobs, it seemed to Cecil that he might as well haul some out for himself. One good boulder brought in some cash, and it was enough to give him the jade "bug." As well as prospecting for alluvial boulders, he went hunting for the *in situ* material and found a deposit at the head of Jim Creek, close to the 2,000-metre elevation in the Shulaps Range west of the Yalakom River. Although he held a dozen claims on the mountain that included a number of *in situ* deposits, Cece always said that the best material he found was in the form of loose blocks in the talus. He discovered a 30-ton beauty on his Cap #12 property, but it was in an unfortunate location for cutting and drilling because of the lack of water, so the quality remains unknown. The material is likely still there.

Cece produced only a small amount from his claims, and there was still an unknown reserve. Jade was found associated either with rodingite or chert on Cap #1 and Cap #3, but it appeared as isolated inclusions in the serpentine and not part of a continuous structure. Elsewhere, the jade was sandwiched between greywacke on the west and black serpentinite on the east. Cece's Jim Creek jade was an apple green colour and highly prized.

While much of the jade found in the province was sold overseas, there are a few famous pieces that have stayed home. Of special interest is the jade boulder at the heart of Arthur Erickson's design for Simon Fraser University in the City of Burnaby/Vancouver. While casting around for a suitable focal piece to be used at the university's official opening in 1965, the idea of a jade stone was proposed by Professor Allan Cunningham, head of the History Department. As jade boulders "don't grow on trees," the question of money was only settled when the O'Keefe Brewing Company offered to cover the costs.

Peter White, an O'Keefe employee and rockhound, was put in charge of the project. He knew where to go and in due course contacted most of the Lillooet dealers and then visited the area to inspect their wares. He subsequently submitted a report on his findings to the university president, Dr. P. McTaggart-Cowan, who made the final selection: a six-ton boulder from Hell Creek in the Shulaps Range.

The claim was held by a Mr. Forster and the chosen boulder was moved to Chandlers Memorial Works in Vancouver, where a polish was added to the natural stone. Twelve thousand pounds of jade at a couple of dollars per pound, plus haulage and polishing charges, probably translated into a lot of beer sales for O'Keefe. The boulder was subsequently used at the opening ceremonies in September 1965, after which it was put in storage for two years while academics, boards, and architects struggled to agree on what to do with it. Given its unusual size and shape, it finally found a home in the very centre of the academic quadrangle, in the middle of the pond. A newspaper report at the time suggested that the O'Keefe Brewing Company paid about $1,000 but noted that the boulder might be worth as much as $100,000 if the surface quality carried all the way through.[7]

In January 2005 the jade fields in the Shulaps Mountains were no longer active. It is unlikely that all nephrite has been found and mined, so further exploitation will likely occur. It is odd that the area is being neglected when the resource has barely been touched, and the location is so much more convenient than others in the centre and north of the province. A possible explanation could be that the latter two areas now have almost 40 years of exploration and delineation, with their reserves known and their quality determined.

The Omineca rush

At the start of the jade rush the Lillooet district had certain advantages, owing to its proximity to railways and highways and to the main centre of economic activity, the city of Vancouver. However, by the late '60s the region was getting crowded, and some prospectors began looking north for other jade deposits. In 1962 J.B. Thurber and H.M. Kindrat found nephrite on Vital Creek, 40 kilometres northeast of Takla Landing in the Omineca. They hauled out three large boulders, weighing 2.2, 0.9, and 0.3 tons. These were reported found about 800 metres upstream of the road bridge.

Subsequent discoveries were made on Quartz Creek in the same area. Then, in 1963, Bruce Russell of Fort St. James found alluvial jade on his placer lease on Kwanika Creek. Many tons of boulders were brought out—some weighed as much as eight tons. Another group of prospectors found jade on Van Decker and Baptiste creeks, in the Mount Sidney Williams area.

In 1963 Larry Owen, a school principal, and his family moved from Los Angeles, California, to Manson Creek, B.C., "buying the town" as his wife Margaret later put it in the title of her book about their move. When they'd left California, they were still unclear where they were heading. "B.C. or Alaska," Margaret later recalled; they hadn't much cared which at the start. As it turned out, the family ended up in Manson Creek because they couldn't resist the cheap price of real estate. There were 18 inhabitants in the town at the time.

Larry was a romantic who dreamed of hunting, fishing, and prospecting and who subsequently put Ogden Mountain on the world jade map. In 1967 he and his future son-in-law, Stan Porayko, found jade boulders in Ogden Creek, northwest of Larry's home base at Manson Creek. Two years of summer mining followed, but always there was the question of where the source was. In July 1969, while out prospecting, they sat down for lunch on a large, white rock. Slowly, over the time it took to eat a few sandwiches and smoke a cigarette, they realized they were sitting on the *in situ* deposit.[8] "There was an 85-foot rock face, and all of it jade," Larry later told reporters. "It wasn't granite. It wasn't a serpentine outcropping, and it wasn't like anything else we'd ever seen before."

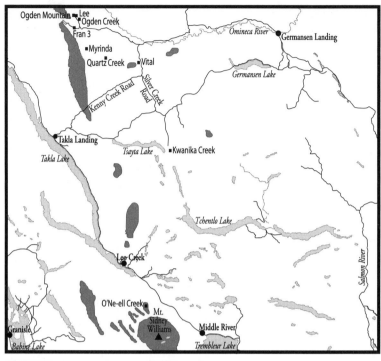

Map 2: *The remote and rugged Omineca jade discoveries followed from the earlier Lillooet finds. The region's mining activities centred on O'Ne-ell Creek and Mount Ogden.*

"It sure looks jadey," Stan Porayko commented.

"It sure as *hell* looks jadey," Larry replied.[9]

Initially, the venture did well for Larry. He had found the motherlode for the many jade boulders that had turned up in streams draining Mount Ogden. The source was so big that a specially designed wire saw was built and shipped in to cut the boulders into manageable chunks, so the smaller diamond saws could reduce them further. The venture was incorporated under the name of TG Explorations. Some of the material was as good as any ever found in British Columbia, and Larry Owen became an avid jade miner and developer.

After the *in situ* discoveries, the next step was to find a location in Vancouver with warehousing and shipping facilities. However, getting the material down to the coast was expensive. It needed to be cut and graded, and Ogden Mountain was a poor base for operations during northern winters. Besides, most of the potential customers were from China and Taiwan, so the Port of Vancouver seemed the only sensible place for a cutting plant. Like all remote northern operators, TG Explorations was discovering that the cost of maintaining a camp in the north and a base operation in Vancouver, plus mining and transportation, soon mounted so high that new working capital was needed.

Two Calgary-based businessmen, David Saxby and Gary Gallelli, were brought in to organize a new company called, appropriately enough, New World Jade. Dave Saxby, who had an engineering degree from the University of Alberta and an MBA from Harvard, came on-board, in charge of corporate development. Within a few weeks he was flying off to find buyers. "We found that other companies had promised a quantity and quality they couldn't deliver," Dave later said. "So B.C. had a poor reputation, and we had to work hard to establish our credibility."

Not speaking any Chinese, he closed his first deal for a substantial order in some confusion, then rushed back to Vancouver to organize the cutting and shipment of 34 tons of jade. There were no suitable saws available, but the company tracked down Al Hampson, who had been in the stone-cutting business since "Mt. Logan was a molehill." New saws were rapidly designed and built and the order filled. Dave made a return visit to China when the order was delivered to ensure that the buyers were happy. "They only tell you what they want you to know," he recalls. "But I knew they were interested, because they got excited when they looked at our jade." The Chinese placed another order. They also got him a front-row seat to a ping-pong match, for which the locals lined up around the block to buy tickets!

The Jade Queen

Once a mineral deposit or placer creek has been found, people wonder how it was first overlooked. The truth is, there are a lot of creeks, ridges, and plateaus in B.C., and prospecting them is a slow and painful business.

Bushes and trees cover critical clues, gorges are often so deep that they can't be climbed down into to sample, and then there's the weather—or rather, the seasons. Summers are short and fly-plagued; winters are long and slow to leave the high country. Yet sometimes the clues are all there. It just takes someone to put them together, so that afterward everyone else can say, "Well, that's obvious." Such was the case with Mount Sidney Williams.

In the *B.C. Minister of Mines and Petroleum Resources Annual Report* of 1960 (a full eight years before the Jade Queen discovery), the already commercialized ultramafic rocks and serpentine deposits of the Letain Creek area were described in detail.[10] Immediately following the geologist's report was a brief note on Mount Sidney Williams. Although the two areas are separated by 450 kilometres, the geologist used almost identical language to describe them. Yet for eight years no one drew the obvious conclusion: if there was jade at Letain Lake, and Mount Sidney Williams had similar geology, then there should be jade there too.

Win Robertson was in Yellowknife on a rockhounding trip in the early 1960s. She camped at the local municipal site, close to a group of University of Alberta students on a geology trip who were making a lot of noise early one morning, tripping over Win's guy lines and generally making it difficult to sleep. The smell of fresh java drifted across, so Win leaned out of her tent and shouted, "If you're walking over me, how about some coffee?"

In due course, a mug appeared. Win extended her hand to get it, and a voice said excitedly, "Where did you get that ring?" On her finger was a jade ring she'd made from a find in the Bridge River area. "I made it," she replied. Her questioner turned out to be Dr. Robert Folinsbee, jade authority and university professor, and it was the start of a long friendship that would lead, ultimately, to the discovery of the Jade Queen property on O'Ne-ell Creek.

In the man's world of prospectors and prospecting, Ms. Winnie Robertson made some good copy for journalists. In 1968 her activities in the Takla Lake area of Central British Columbia hit the papers. Win was used to moving house every few months and starting all over, because she was married to a civil engineer. Transferred to Edmonton in 1962, she again met Dr. Robert Folinsbee, and it was then that Win became fascinated with jade, studying it in-depth. Later, when the Robertson family ended up in Prince George, she found herself at the gateway to the as-yet unproven jade fields of the Omineca region. When Win made her first jade discovery, the newspapers of the day gave the impression that she was just a lucky housewife who "happened" to find jade.[11] The newspapers got it wrong: Win Robertson had done her homework. After the years studying jade in Edmonton, the move to Prince George in 1966 allowed Win and husband Stuart to connect with local jade miners like Art Bell, Larry Owen, and Bruce Russell. They made trips into Kwanika Creek, with Dr. Folinsbee joining them on one of their explorations. Win wrote:

> After studying the GSC maps and reports, I decided (with the help of Dr. Folinsbee) that if the scanty information on the formation of nephrite was correct, then it would have formed where the soda granites intruded into serpentine, altering tremolite and actinolite into nephrite at the contact zone. The Mount Sidney Williams area of the Middle River between Takla and Trembleur lakes, where Armstrong had mapped a serpentine belt with a soda granite intrusive in 1937, should produce nephrite.[12]

This was no casual rockhounding exercise, but a carefully planned geological exploration.

In 1968 Win and new partner Denis Oram, a jade "addict" and manager of the Bank of Montreal, Prince George branch, formed Tezzeron Nephrite (named for Tezzeron Lake nearby) with the idea of exploring the creeks in the Mount Sidney Williams area. Since the region was impassable on foot, they chartered a helicopter. In three days, Win and her 16-year old son Gordon found enough jade in the lower section of O'Ne-ell Creek to warrant bringing in a camp. Using pack drills, they bored a number of boulders to test the quality and threaded chokers through the holes for chopper-slinging. However, something puzzled Winnie. None of the boulders were in the creekbed itself; instead, they lay on a wide gravel flood plain on the north side. Where had they come from? Judging by the moss growth, she estimated they must have been there 20 to 25 years at least, possibly brought down during the big 1948 floods.

The other curiosity, as they worked their way upriver, was that the boulders became larger and larger and were always water-worn and on the north side. Win suspected there must be some sort of barrier higher up and that only exceptionally high water moved the boulders down. Farther upstream the valley narrowed and was then blocked by a 30-metre-high waterfall. The side cliffs were covered in dense alder and devil's club, and it took considerable effort to reach the top of the headwall. But once there, the mystery was revealed. On the north side of the valley was a rock ridge about 2 metres high and 10 metres long, with huge boulders piled up against it. Most were jade. The lower surfaces were polished, water-worn, and easily identifiable. The boulders out of the river were a ragged grey colour and not obviously nephrite.

In her own words, Win Robertson was "stunned." All they could see were jade boulders, many 10 to 15 tons in weight, piled in chaotic abandon for over 200 metres and disappearing upstream. "I never imagined I would locate the deposit I did," she would later write. " ... It went beyond anything anyone could imagine and was exactly where Armstrong had mapped the intrusive. I kept telling myself, 'Winnie, old girl, this doesn't exist.'"[13]

The contact zone ran for about 800 metres up the mountain and was about 400 metres wide. The party stayed in the lower area for another week, readying material for an airlift. The upper zone was too difficult to reach from the lower camp. Win called Denis Oram on the radio phone, but mindful that anything they said could be heard by others, they talked in a pre-arranged code. "Did you find the pony in the pile?" asked Denis. "It's a darned Clydesdale!" Win replied. "You'd better get up here."[14]

In late August they cleared a site above the waterfall and moved the camp. Mice arrived from nowhere and almost ate them out of food until everything edible could be stored in metal drums. They drilled the smaller boulders (less than two tons) and readied them for air haul. A heavy-lift Sikorsky S-58 was chartered. With its capacity of 3,500 pounds, they were limited in what they could remove without a saw. While this was going on, Win staked 40 claims along the creek, calling them Genesis 1–40. The name seemed appropriate: it was a new beginning.

In mid-September 40 tons were slung out via three-minute helicopter rides to a barge on the Middle River for transport to Fort St James. From there, Canadian Freightways back-hauled it south to Vancouver. Exhausted, the party flew out on the S-58 to Smithers. Win Robertson had spent 68 days at O'Ne-ell Creek—only eight of them without rain. When she arrived in town she was looking, as she later put it, "less than glamorous." As she walked into the pub she was toasted by the prospectors present as the "Jade Queen." The name of the mine was born.

With that first shipment from O'Ne-ell Creek, the chopper slung an 1,800-pound boulder down the valley, using a chain threaded through a hole bored through the middle of the rock. Both the core sample and the boulder's skin looked beautiful, and Don Lee, who was marketing much of the jade to eastern buyers at that time, offered Win three dollars a pound for it. Win turned him down, saying it was worth a lot more. Then she took the next step and had it cut. Years later, Win laughed as she recounted what happened next. "It was proof that you can put a drill right through a boulder and hit the only decent bits in a boulder of garbage! I should have taken Don's offer at the start."[15]

Win was unprepared for the media frenzy that followed. Newspapers, radio, and TV reported outrageous claims: "She has found a mountain of jade!" trumpeted the *Royal Columbian*. "Exotic bids made for jade!" roared the *Province*.[16] *Good Housekeeping* magazine sent a team of reporters from the States to cover the story, complete with the pink apron they wanted Winnie to wear for the photo shoot. Penny Tweedy and colleagues from *The London Illustrated Times* arrived to do an article on the jade mine, so Win took the visitors into O'Ne-ell, only to find that a grizzly bear had been through camp and had caused extensive damage. The side of the trailer had been torn off, the stove ripped out, and the deep freeze battered. The refined Londoners were considerably put out by the damage, but Winnie didn't bat an eye at the chaos. "Probably got into the curry powder," she remarked.[17]

Tall, slim, and leggy, Win had in a previous stage of her life been a fashion ramp model and an interviewer for CBC TV's *Ladies First*. By her own admission, she had at times appeared as a "champagne parfait blonde," thanks to weekly ministrations at the studio hairdresser's. The media and the public lapped it up. There were some downsides, however.

Every nut case in the country either wrote me or turned up on my doorstep, thinking I should give them some money, as I was making millions. One man showed up who had quit his job as a lighthouse keeper on the Great Lakes. He was going to marry me and protect me from all the unscrupulous men who would try and take all my money. I told him to stick around until my husband got in from work and to talk it over with him.

Another day, a group of "Born Agains" expected me to give them the money to build a church somewhere in Georgia. Their theory was that I must be a God-fearing Christian, for the Lord to reward me by leading me to "The Jade Mountain." Obviously I had prayed and called on the Lord for help! I did use some of those words, but not quite in the order they had in mind.

Inevitably, Tezzeron needed capital to expand, and this it did by forming Jade Queen Mines Ltd., with Win's group holding 49 percent, and Columbia Athabasca Resources (a subsidiary of Sandwell Co.) the balance of 51 percent. The latter sold some of its shares later to Aasamara Oil of Calgary.

Win had to adjust to her new-found celebrity, as she was often recognized in the street. People approached her asking if they could invest in "her jade mine." They had no appreciation of the complexity and uncertainty of the jade business, where 97 percent of the material remained on the slopes where it was found simply because it wasn't worth the price of moving it out. But being well known had its advantages.

In 1969, when Win and the Sandwell executives were negotiating a new contract, it became apparent that Sandwell, as did financiers, believed they had the upper hand and were able to dictate terms to the Jade Queen. Win's lawyer suggested she call their bluff and talk to her own bank. She walked into the local branch in White Rock and asked the manager how much the bank would be prepared to forward. She walked out with a $50,000 line of credit—an enormous sum of money in 1969.

Within days, Win had mobilized her field crew, chartered a helicopter, ordered camp provisions and flown into O'Ne-ell Creek to start the season's program in May. Some days later she was in camp when she received a message that Sandwell urgently wanted to talk to her. Hastily grabbing the helicopter out, she was hit with the realization that although she had a huge line of credit, she was also personally liable for that same amount. The thought of owing that much money terrified her, and she broke out in a sweat. At Smithers the helicopter was late for the only scheduled flight south that day, so the pilot landed on the runway, effectively blocking the outgoing plane from taking off until Win had transferred to it. There were no seats left so she sat up next to the pilot, who knew her by reputation. "Win," he said, "you smell!" In Vancouver she had a chance to wash, change and meet with Sandwell, who agreed to her terms.

Under the new structure, Win Robertson was president and in charge of field operations. Roy Hinken from Sandwell was appointed to run the finance and office end of the business. A Vancouver-based plant was set up under Bill Yarmack, and jade was cut, sold and shipped to Asia and Germany. Included in this output were jade coffee tables made by Lorne Marshall for B.C. House at Expo '70 in Sapporo, Japan. Another successful miner, Egil Lorntzen, who had developed the huge

Win Robertson poses for a glamour magazine (that brought the snowshoes specifically for the photo) after being hailed as the "Jade Queen."

copper mine that would become Lornex near Logan Lake, B.C., took a shine to Jade Queen products. He had the doors on his new mansion made with brass plates and jade insets, and the swimming pool lined with jade tiles. This made another "splash" in the media and was good for creating public awareness.

Back in the field, the sheer remoteness of the site made the 1969 season a challenge. Expectations were high after a winter of furious financing and corporate jockeying. The helicopter platform that had so laboriously been constructed in the upper basin of O'Ne-ell Creek had been crushed by spring runoff, and it took a while before jade could even be lifted out. The pilot in charge was "Timber" Tom Lansley. An early foray into the valley showed the creek to be too full for any meaningful work to be done until the water level dropped. Special saws were also designed, and a high-pressure hose system was developed to wash the overburden off the contact line. Win worked with Mike Bealey, doing further assessment work.

However, the costs of helicoptering equipment in and freight out versus road transport at the Mount Ogden and Kutcho Creek jade sites made the recovery uncompetitive.

In 1970 the reserves on the Jade Queen properties were estimated at eight million pounds,[18] and the company continued to stake around Mount Ogden in ensuing seasons.[19] Harv Evans was the chatty pilot who did some of the load hauling, and his flight engineer was Tom Lansley. It was dangerous work, but nothing seemed to faze Harv as the big blades swept the air, often perilously close to trees in the tight confines of O'Ne-ell Creek. In 1970 "business wrangling" resulted in Win selling her interest in the company. There were accusations of fraud over whether drill cores had been tampered with, and late in the year Win stepped down, leaving CEO Roy Hinken in control of the company. Compounding the problems in upper management, the price of jade fell. When Jade Queen terminated Win's contract, in her words, "the only person who made money … was my lawyer attempting to sort through the tangled maze of Athabasca's dealings."[20]

Today Win Robertson remains active in the lapidary field and is a past president of the B.C. Lapidary and Mineral Society—the umbrella organization for the 30 clubs in the province. The Society publishes a quarterly magazine called *The British Columbia Rockhounder*, of which she is the founder and editor. In 1972 geology student John R. Fraser completed a university thesis covering the geology of the Jade Queen deposits in considerable detail, although the question of reserves was not answered.[21]

Production ceased in 1970, after amalgamation with Mount Ogden, the latter being less expensive to operate. The claims came open in 1983 and were staked by Lorne Warren of Smithers. However, the following year, when he went to work on the site, the B.C. Department of Fisheries closed him down. Apparently O'Ne-ell Creek is the northernmost limit of the Fraser River spawning system and will never be mined.

Mount Ogden is a broad upland with room for more than one producer. Next to David Saxby and Gary Gallelli's New World Jade, another company owned by Howard Lo and Larry Barr, called Nephro-Jade Ltd., went into production. To the east of New World Jade was a property first staked in 1968 by Ed Tucker and later optioned to Dr. Glen Kong of Kuan-Yin Industries Ltd. It was the company's intention to start producing jade items locally, in competition with the cottage industry in Taiwan. Kuan-Yin Industries was a private company incorporated on April 8, 1969, that went public in August of the same year. A month earlier the company had taken over the assets of Northern Jadex, a private company with jade claims on Mount Ogden, adjacent to New World Jade. Much of the financing was provided by Dr. Glen Kong and his associates. They brought out nephrite from the claims and set up a shop in North Vancouver to manufacture various jade items. Despite the public offering, Northern Jadex did not prosper and was eventually taken over by New World Jade.

The Mount Ogden properties weren't the only show on the hill. About four kilometres south as the crow flies, the Fran 3 occurrence was developed by Neil Scafe, owner of the claim, and Lorne Warren. The property consisted of a nephrite lens within the Cache Creek ultramafic unit. Jade occurred along the contact between serpentinite and structurally overlying metachert and greenstone. The owners extracted about 90 tons of low-quality nephrite from the locality in the mid-1980s.[22]

A newspaper account in the *Vancouver Sun* of January 24, 1970, was headlined: "In B.C. Big Jade Discovery Disclosed." Jade mining has always made good copy for journalists, who are often unaware of geological realities. Articles on the potential growth for jade are written in the same way as articles on gold or metal ore bodies, but the market is not there. Production numbers are hyped—or, worse, extrapolated.

A case in point was the information written about an event that took place in 1970, when an 18-ton boulder of jade was shipped to Japan for exhibition at the World's Fair in Osaka.[23] It was insured for a million dollars. Simple math translates this as $25 a pound. While it is possible at times for the price of small blocks to reach this value, it's not true of large boulders, where there is the uncertainty of quality throughout the material, plus considerable expenses involved in reducing the mass to manageable and saleable dimensions. Nevertheless, the media referred to "the million dollar boulder" frequently.

Pacific Jade Industries succeeded New World Jade in the Mount Ogden location, and the company issued shares to bring in capital for expansion. Their 1972 production amounted to $200,000, half of which was exported as raw jade, with the balance being turned over to local Canadian artists. By 1973 the company was well-established in saleable art, and in the early part of that year a sale of finished pieces in Calgary had returns of $80,000. However, in the same year both the mineral property and the facilities in Vancouver were sold to a Japanese consortium operating as Continental Jade Ltd. The new business was managed by a retired Japanese military general, Mr. Yamamura.

Around the same time, Oscar Messerer of B.C. Gem Supply in Lillooet had an interest in Ogden Mountain leases, but not for long. The jade industry was going through a series of booms and busts, which drove takeover businesses into a frenzy. Properties changed hands often, and money was made and lost as newcomers learned quickly that hidden expenses could bankrupt an operation in a season or that a windfall rise in price could allow new properties to be bought up and operated with the associated economies of scale. But gradually, as the '70s passed, there was a convergence of sorts. In time, Continental Jade's assets were acquired by Joe Bell and his company, Jade World. Joe operated the mine with the help of a small crew.

The treasure of Mount Ogden continued to supply the raw material for export, mainly to Taiwan, but, increasingly, to the People's Republic of China and on a continued small-sales basis to the U.S., Germany, and within Canada. After Mr. Bell's death, the mine was continued by his business partner, Kirk Makepeace, under the new name of Jade West, and with that change came a period of stability. Jade West holds and operates the property to this day. Indeed, the company has become the principal producer of jade in Canada and hence in the world. The market is not large, as only a few hundred tons are needed annually, and while this could be supplied by either Siberia or Australia, neither country has been a serious competitor to Canadian producers to date.

Dease Lake and Cassiar country

Some two hundred million years ago, when there was no Dease Lake (nor indeed a province of British Columbia), the region where it is today was below sea level. A series of tectonic plates, or terranes, carrying volcanic island chains and sedimentary rocks moved northeast on the giant conveyor belt of continental drift and ground into the North America Plate, which at that time was considerably smaller. This terrane, called Stikinia by modern geologists, bent and buckled and built up the continent's western edge in the form we see today. Later, ice ages—there may have been many, but certainly at least three in the past two million years—smoothed the peaks and troughs, softening the landscape. Still later, the Tahltan peoples reached Dease Lake after their ancestors crossed the Bering Strait and moved down into a game-rich but people-poor North American continent.

In 1834 the first white man to see Dease Lake was Hudson's Bay

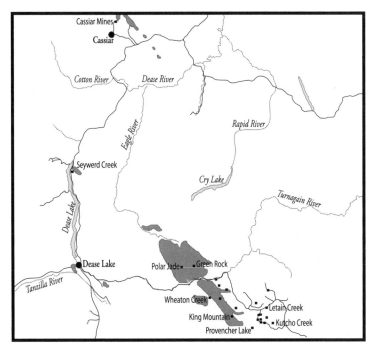

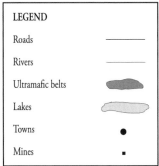

Map 3: *The early discovery of gold at Dease Lake, and later of asbestos (a close relative of nephrite) at Cassiar, opened up the region. Today most of the province's jade production comes from the Kutcho Creek region east of Dease Lake.*

Company employee John McLeod, who named the 40-kilometre-long body of water for chief factor John Warren Dease. Gold was discovered in 1874 and sporadically thereafter, but the area has always attracted, and been populated by, people who sought wealth under the ground. When the jade boom started in the early 1970s, a remark I made that Dease Lake was getting to be the jade capital of the world was taken seriously. In short order, prospector Steve Simonovich erected a sign to that effect, and a large nephrite boulder was placed at the entrance to town. Thereafter, it was a rare day when you didn't see a tourist sitting on top of that rock, having his or her photo taken.

With the increase in jade mining activity, the Geological Survey of Canada began to delineate deposits associated with the ultramafic belts. There were GSC camps at both Dease and Turnagain lakes in the 1970s. The latter lake was named by the great explorer Samuel Black, who ended his exploration there in 1924. Many of the non-Native inhabitants had come to the region for the gold, and between 1880 and 1940 there had been steady action, albeit on a small scale. When the gold mining dried up, these men turned their hands to jade prospecting. Most finds were fractured, had inclusions, or were of poor colour, but there were enough solid boulders to keep the prospectors in a grubstake, although funds often disappeared as quickly as they were acquired—usually in the bar. Times were quite different for those who followed after the Second World War. There were helicopters and portable radio telephones, so even in camp men had contact with the outside world, and there were luxuries such as fresh food. In Dease Lake there was television.

The Cassiar Mountains form a long belt of granitic and intruded rocks, part of a great batholith lying west of the Rocky Mountain Trench (now filled by the W.A.C. Bennett Dam) and east of the Stikine Plateau. The mountains are about 150 kilometres wide and extend from the Yukon border in the north to the Omineca Mountains in the south, which they adjoin. They are divided into a number of ranges, of which only the Stikines can be considered "a jade mountain range." Some of the peaks reach over 3,000 metres in height, and, being in the far north, the timberline is low enough to expose bedrock over much of their elevations. Prospecting for minerals is therefore simplified by the abundance of bedrock. The lack of any superficial cover exposes the serpentine rocks on which vegetation is further inhibited by the chemistry of the soil. As far as is currently known, jade in the Cassiar Mountains extends in a narrow belt about 110 kilometres long, from McDame Mountain and the Cassiar Asbestos Mine in the northwest to Kutcho Creek in the southeast.

Over time, the GSC mapped the three distinct jade areas discovered in the district. East of the village of Dease Lake and centred around

Letain Lake (including Kutcho Creek and Provencher Lake) was a large zone where Mohawk Oil, Nephro-Jade, and later Jade West had their activities. There was a second outcropping at the north end of Dease Lake, extending across the lake from Thibert Creek and along Seywerd Creek to the east. Finally, there was the third jade zone located at the Cassiar Asbestos Mine.

Probably the first person to recognize jade in the Cassiar district was Bill Storie, who, as a placer miner on the Fraser River, knew what nephrite looked like. In 1938 Bill was working his placer gold claims on Wheaton Creek in the Dease Lake area and identified jade boulders on his northern leases. The valley had escaped heavy glacial scouring and was generously sprinkled with large boulders from the upper slopes. However, there was little value in the stone at that time, and when the war started in 1939 Bill enlisted in the army and his claims lapsed. After the war he came back to the area and lived in Cassiar townsite, a convenient location from which to work his renewed gold claims nearby.

About the same time, several other prospectors moved into Wheaton Creek. One of the first was Gerry Davis, who knew about jade. Early in his life he'd been a pilot on a Mackenzie River boat sailing from Fort Nelson to Aklavik, and later a provincial policeman working out of Telegraph Cove, where he'd kept a dog team for getting around in the winter. As part of his job as an officer, he knew everything that was going on in the northwest corner of the province, and that included what all the prospectors and placer miners were up to. He learned from them, then staked some claims on McDame Mountain, which he later sold to Cassiar Asbestos Ltd. With the money from the sale he took flying lessons, bought himself a plane and was still flying at 90 years of age. For many years, when Gerry was a lot younger, he camped at the foot of Wheaton Creek in a comfortable cabin with his wife Agnes, who was there for the summer. Agnes fed the birds and chipmunks and was always glad to see any passersby at the cabin. Helicopter pilots dropped in frequently for tea and cookies, and she always had a supply of tobacco for any addicts who had run out.

When the gold claims on Wheaton Creek came due in 1956, Gerry had them registered. He then began cutting up the jade boulders on his placer claims, flying out the production in his Piper Supercub. In 1957 he shipped 1,000 pounds, and in 1965, at the height of production, he managed almost 10 tons. Both the gold and the jade were of interest to mining companies, and at one time Gerry optioned them to Dempsey Mines Ltd. A report in *The Province* in 1971 notes:

> An immediate start in jade production is planned by Dempsey Mines Ltd. at its placer leases on Wheaton Creek, 45 miles east of South Dease, on the Cassiar–Stewart highway.
>
> From the proceeds of a $100,000 underwriting, $70,000 has been used to acquire three key leases; and the remaining $30,000 will be used to cut and ship jade from the property, which contains a total of 16 leases.
>
> Consulting engineer Peter Sevensma has reported a minimum of 300 tons of jade in sight, with an estimated market value of $2 to $2.50 a pound. Two diamond cutting saws are on the property.[24]

Dempsey Mines did a little work on the leases, but eventually they reverted back to Gerry, who sold them to a Vancouver-based consortium in 1983. Gerry dropped by to visit me one night around that time, and rather than bringing a bottle of wine, he brought a poke of gold and set out enough nuggets so that my wife and daughters could each have a pair of earrings! The following morning he reappeared at my office and called out: "I just remembered you have a third daughter, so here's a set for her!" Gerry and Agnes were very much a team, and after Agnes died in 1984 Gerry wrote, " … we had just about 43 good, close, and adventurous years together, in which we saw nearly all the countries of the world north of the equator."[25]

Andy Jensen, an eastern prospector who came west to try his hand in the placer fields, worked east of Gerry's properties. Andy had heard about jade, and when he found a boulder was bitten by the jade bug too. Gerry helpfully advised him to stake the area. Production came from alluvial boulders along the valley of Provencher Lake (named after Conrad Provencher, an early prospector), and in the creek leading out to Wolverine Lake. Andy staked other claims in the area as well and enlarged his holdings to the point where they became too much for him to handle by himself. Andy formed Nephrite Mines Ltd. to develop his discoveries in Provencher Lake, and in 1973 he leased those claims to Nephro-Jade Canada Ltd, which began the job of unearthing the alluvial deposits in the vicinity of the lake.

Walter Ellert showed up in the area about the time that jade was gaining prominence in the mineral industry, and he acquired some properties at the head of Wheaton Creek. I visited the claims with him and agreed to make a short side trip to sample a jade outcrop at a remote location. Walter arranged for Carl Simmons to helicopter us there, along with some equipment from Walter's Wheaton Creek claims.

The unproven outcrop was a few miles east. The Hiller 12C chopper was grossly overloaded with food, equipment, and a diamond saw and landed very hard in a downdraft. You need to appreciate that at 6,000 feet, the helicopter had very limited lifting power. Carl was initially pretty shaken up, as he knew better than the rest of us just how close we had come to "piling in." He sat there for half an hour recovering his nerve. Then he became worried that they'd damaged the chopper, but it had survived, and he duly left. However, problems cropped up for the other two after setting up the diamond saw and firing up the gas motor: the spark plug blew out of the head, leaving Walter and me without cutting ability or power for the next three days, when Carl was due to return.

Near Provencher Lake, at the 2,000-metre level, a vein of nephrite zigzagged across the slope for about eight kilometres and was known as the "China Wall." It was from this lode that many of the alluvial boulders had weathered and fallen, according to Jack Gillies of CanJade.[26] About the same time, Hal Komish from Watson Lake staked the Barb Lens, which I had first seen in 1976 with Walter Ellert during a mechanically challenged few days of camping. We had heard the sound of a diamond drill in the cirque to the west of camp and had walked over to see what was going on. Two men were drilling talus blocks of jade that were strewn around a small rock bowl. The blocks obviously came from the lode above, which was later called the Barb Lens. I had seen the indicators of this lode the year before from a point a mile away on the other side of the valley and had predicted its existence. The lens, upon inspection, indicated reserves of about 3,000 tons, and at one time, before erosion had spalled off the surface blocks, it must have been a very large deposit—certainly the largest seen at that time in B.C.

Another early prospector in the area was Ben Seywerd, who, with his brother Josef, found jade at the north end of Dease Lake, along what was later called Seywerd Creek. Across the lake on Thibert Creek, he uncovered more jade.[27] It was in July of 1963 that the first alluvial boulders were found, and two years later the *in-situ* deposits came to light. Ben and his brother had come to the region via Wheaton Creek, where, strangely, they had not seen jade. They may have found the entire area previously staked. Josef Seywerd already was a seasoned jade prospector and knew how to sell the product as well; in the early '60s he had tackled a 330-ton block of jade in the Lillooet area and had exported material to Germany and other countries.

Walter Ellert later acquired Ben Seywerd's claims at the north end of Dease Lake after Ben had decided the long trip north was not as rewarding as it had been when he was younger. There was a small amount of asbestos showing on the claims, and Walter was able to interest Cassiar Asbestos in taking a look, but their geologist didn't think much of the possibilities. The claims were returned to Walter, with a good many years' assessment work recorded against the claims. This meant that Walter had control of the claims without the need for further work for some time.

Under the terms of the Mining Act, in order to hold a claim, the claim holder must do a minimum amount of work on the property or pay a fee each year to the B.C. Department of Mines. Since most prospectors are long on energy but short on cash, they usually opt to perform work, such as brush clearing, road upgrades, draining, ditching, surveying, chemical analysis, or mining, rather than part with hard coin.

In Wheaton Creek, both Walter Ellert and Andy Jensen had a few tons of jade boulders gathered up. They contracted Grant Stuart of Watson Lake to take a caterpillar train in to haul the stockpile out one winter. It took several D7 Cats and sleighs, plus the use of a bunkhouse and cook shack—it was mid-winter and as bitterly cold as only the north can be. The job cost $50,000, and they couldn't pay the bill, so the jade was trucked to Watson Lake, where it remained under Stuart's watchful eye. Eventually Walter paid off his part of the bill and recovered his jade, but it was quite awhile before the account was settled.

Frank Plut was a bulldozer operator and catskinner and another of the Dease Lake locals. Hauling other people's boulders from Mount Ogden in the Omineca, he decided that he should get into the jade business himself. So in the Cassiar country he hauled even more jade, staked claims and got into sales. Before long other prospectors moved into the area, which then included Provencher Lake, Kutcho Creek, Dease Lake, and Cassiar. Frank Plut, along with Steve Simonovich and Don Lambert, were active, and with the increase in players there was a steady ramping up of competition between operators. This was further exacerbated by the way claims were registered.

Jade properties are covered by a "lode claim," and gold claims are covered by a "placer claim." In many locations it was possible for two separate claim holders to stake out the same piece of ground, with all

At one of Walter Ellert's alpine prospects near Provencher Lake, trees are used to construct a trestle bed over an in situ jade boulder. A portable diamond saw then cuts the nephrite into manageably sized blocks, without the expense of first being moved. Walter is on the right.

the associated friction that you find whenever two personalities have to share anything. One operator would move rocks off his area of interest and pile them up where the alternate claim holder was working or was just about to work. There were arguments, fights, and even death threats at times. The north attracts strong personalities, and many choose the solitary prospector's life with good reason.

Being a mineral commodity, jade has come under the scrutiny of mining promoters, with the view that capital could be raised for the development of a property, just as is done for any other mineral deposit such as gold, copper, or iron ore. There should be no stigma attached to the word "promotion," although there have been abuses of the process. Insofar as the scale of the jade business precludes large profits, however, the promotion of most jade properties is scarcely justified. There is no need for great sums of money in order to mine and produce jade. No complex milling process is involved, and basic equipment amounts to a bulldozer or a backhoe, a few diamond saws, a small camp in the field, and a couple of workers. The material is cut into manageable pieces of 50 kilograms or less, and the transportation can be contracted to helicopter or trucking companies. It is, in fact, an ideal mining venture for the individual. Nevertheless, some mining companies have tried to promote jade properties, with few having more than initial success.

Apart from Cassiar Mines, which was primarily interested in asbestos, not nephrite, all the regional players were small, underfinanced, and limited in what they could do individually. In 1975 that was about to change.

Mohawk Oil was then a small, privately owned company with a string of gas stations based initially in Vernon, B.C. Their chief executive officer, Don Skagen, liked the idea of getting into the hardrock mining business, although the company had no experience in exploration or development in either hard rock (ore) or soft rock (oil and gas). However, being better financed than many of the small-time prospectors, they bought up some of the Kutcho Creek properties and consolidated the Dease Lake nephrite mining to some degree. Being privately owned, they had no access to large sources of risk capital and so were acutely aware that they needed to ensure sales before developing. They took over some of Andy Jensen's, Hal Komish's, and Walter Ellert's claims and staked others to form Cry Lake Jade Mines Ltd. They then set up a camp and built an airstrip at Kutcho Creek to service their new, consolidated mining operation.

Via the Lees in Langley, the company was introduced to two buyers from the Orient, whom they flew into the Cry Lake valley. Standing atop a 27-ton jade boulder, negotiations were started with the potential clients. In order for the Chinese buyers to be assured of material, Don Skagen wanted them to put some money up front—sufficient to buy two D-6 Cats at least. The buyers were sympathetic, but there was a problem: they did not know the Mohawk Oil crowd, and the Chinese way is to deal only with friends. To solve this problem, both parties agreed that the upfront money would be held by a third party whom both groups knew and trusted—Don and Gwen Lee. Don, experienced with earth-moving machines, was able to find two Cats with the specialized attachments needed for the project at a good price and in short order. Although the deal was struck in September, there was such a quick turnaround that the machines were in the field that fall, and production was possible that same year. Based on that start-up, Don Lee became Mohawk's jade agent, and thereafter he and Gwen would fly into Cry Lake/Kutcho two or three times a year to grade nephrite at the strip prior to the high quality material being flown out.

Under the terms of the contract, Mohawk negotiated with buyers, mainly in Taiwan, to supply 350 tons annually, at three dollars a pound, which translated into over $2 million in annual revenues. Cry Lake used two-tracked Nodwell vehicles, capable of hauling 20 tons at a time, to remove the overburden of soil and glacial till and to haul out the nephrite, at a cost of about 50 cents per pound. With the Canadian material being cheaper (and probably better) than the Taiwanese product, a good market developed. There were also sales in other countries, such as mainland China, the U.S., and Germany. In 1979, when the jade market began to decline due to overproduction, Mohawk attempted to stabilize the industry by acquiring first rights to buy jade from Cassiar Mines. In this way, if they felt Cassiar was dumping jade because of their low mine production costs, Mohawk could buy and hold it to prop up the price.

As in all mining camps, liquor was not permitted on site at Kutcho Creek. As Don Skagen put it, "You wouldn't get anything out of the men if you did." But there were other shenanigans that helped the men pass the time. Don was seldom in camp, but when he was, being management, he enjoyed a few drinks on the quiet. One night after he had partied heavily, the lads stole into his hut and quietly lifted his bed, with him in it, and carried it down to the runway about a hundred metres away. They did the same with Hal Komish, who was another director of Cry Lake. Once there, they placed the two beds out on the strip, arranged a wreath of wild flowers and a cross at each head, and draped tarpaulins over the beds in case it rained. In the morning, Don awoke and was highly amused by the whole incident. Hal Komish was not.

But the good times didn't last long. Under pressure from the Taiwanese buyers to cut costs, Hal Komish considered blasting, as was done in Taiwan. However, the British Columbia Geological Survey (BCGS) geologist advised against it. It was for that very reason that the Taiwanese jade was now so expensive to mine; fractured material from previous blasting had first to be removed by hand—a slow and expensive process.

Instead, the company worked at lowering production costs, notably air transportation expenses, which were high. Early on, the jade was flown out in a de Havilland Twin Otter that had a carrying capacity of just over a ton, but later a short takeoff and landing (STOL) Caribou made the circuit with a full fuel bladder, flying into Toodoggone Gold, located 150 kilometres to the southeast. After discharging its load, the crew rolled up the bladder and flew to Kutcho Creek to take on jade

Surface jade boulders were collected from the surrounding area and stockpiled at Kutcho Creek's airstrip, where they were cut into manageable blocks before being flown out by plane to the roadhead, for trucking to Vancouver.

for the back-haul to Smithers. The Caribou could handle close to four tons, which was more cost-effective, and managed the relatively short 400-metre runway at Cry Lake with ease. Another important factor over which they had no control, but which affected the profitability of the company, was the weather. The usual mining season was extremely short. It started in late June and ran through to September. One year they made it to mid-October, but the limited season put a lot of pressure on the field crew to deal with problems quickly and to minimize downtime.

Bob Yorke-Hardy was a trained mining technologist who came from Smithers and knew some of the Mohawk Oil founders. A prospector since 1975, he'd worked at the big Lornex Copper Mine, prospecting on weekends, and had had a property of his optioned by Mohawk. It was a natural fit for him to take over managing Mohawk's mineral interests in B.C., Alaska, and elsewhere in 1980. Bob realized that although Cry Lake/Kutcho was successfully selling its top 15 percent for jewellery and carvings, there was another 30 percent of lower-grade nephrite that might be turned into construction-grade items, such as tiles. Some years earlier an American, Don Anderson, had tried making tiles from Alaskan jade and had developed the techniques necessary for slabbing nephrite. However, the finishing polish was difficult. Mohawk took over the technology and moved it to Vernon, where Bob spent time intermittently taking the process to a more advanced level.

It was the early '80s, and in San Francisco "Silicon Valley" was ramping up for the computer revolution. Huge numbers of ultra-pure silicon boules were being sliced, polished, and diced to form the basis for the expanding silicon computer chip industry. Bob realized that Mohawk could learn from those electrical engineers. As a result, the company developed a blocking plant that produced four-inch square billets that were about 10 to 12 inches long. These were sent to California, where they were ground square to a perfect 100 millimetres and then sliced off in tiles 3 millimetres (⅛ inch) thick. Being so tough, jade tiles were immensely strong, and the thinness did not make them susceptible to cracking.

The company advertised in a variety of North American "house and home" magazines, and sales were brisk. Despite the initial expense people wanted the luxurious look and feel of jade. Bob shipped a suite of tiles to Australia for a fireplace, and in Virginia a wealthy client had half the floors in her spacious home tiled. The company also sold large amounts of fine jade (selling at $25 a pound) to New Zealand. Aware that the costs were high in San Francisco, Mohawk was close to buying their own machine to do cutting and polishing in-house when the company closed its mining operations in 1981.

In 1980 a Japanese consortium was looking seriously at opening a massive copper-zinc mine in the Kutcho Creek area. This would have included building a road into the area from Dease Lake, 150 kilometres to the west. Presumably this would have greatly benefited Cry Lake's jade operations as well. However, environmental concerns caused delays, and in 1981 Mohawk abandoned its operations. Then in 1984 the Japanese withdrew their support for the potential mine.

In the summer of 1976 Kirk Makepeace took time off from getting his commerce degree at the University of British Columbia and signed on with Nephro-Jade. Those were the expansion years in the jade industry, and Nephro-Jade ran two mining operations: the Provencher Lake property and the Ogden Mountain (Far North) property. Owner Howard

Lo had recently opened up a large market to Taiwan, and demand was strong. Makepeace recalls in a letter:

> The Provencher Lake camp was located in the scenic Cassiar Mountain Range of Northwest B.C. We flew ourselves and all of our supplies in from Watson Lake, Yukon. For $1,500/month we worked seven days a week, drilling the jade boulders that were strewn over the valleys, and hidden under the willows and moss of the pristine valley. Every morning we would pack a diamond drill on our back, along with a water pump and fuel for the day, and then hike up to two miles in all directions from our camp, in search of those boulders. There were no roads for equipment, just the trails of caribou, moose and bear to follow. Our small, temperamental lawnmower engine drills would enable us to retrieve a diamond drill core from the jade boulders. On most days, we would bring home three feet of jade core. If we were lucky enough to drill five feet in a day, we'd be eligible for a bonus of 10 cents an inch. Before the summer ended, I was promoted to mine manager for the Provencher Lake Jade operations, and under the guidance of Larry Barr, of Far North Jade fame, my decision to return to university was postponed, and my career as a jade miner had begun.
>
> During the years 1976 to 1980 at the Provencher Lake operation, our main competitor was Mohawk Oil, operating as Cry Lake Jade Mines. Our operations were vastly different. Under Mohawk's well-financed operation, their workers flew back and forth from their camp to their mining sites by helicopter. They had electricity, a proper camp with showers and toilets, a cook, days off, a pool table, and an annual BBQ with wives and girlfriends flown in for the event. In fact, Mohawk invited all the mining operations in the area for their annual BBQ, with the exception of the Nephro-Jade crew. Theirs was quite a contrast to our tent camp where, in addition to my prospecting duties, I held the job of chief cook and bottle washer. We hiked everywhere, and the only form of transportation was a float-equipped Beaver from Watson Lake, which arrived approximately every two weeks with supplies that included a newly charged battery to power our radio telephone, our link to the outside world.
>
> While Mohawk built an airstrip to fly their jade out to the nearest community of Dease Lake, we would mark our jade boulders with poles from nearby trees, and tie nylon flagging to the top in readiness for Cat train transport. In February, when the ground was good and frozen, we would contract the Cat trains, (a series of D-8 Caterpillar

tractors and trailers). These machines would grind in over the snow nearly 100 miles, dig out our boulders from upward of three metres of snow (that was the reason for the poles with flags on them) and drag them to Dease Lake. It was during one of these winter hauls that I was required to help find some of the boulders where the poles had fallen over. Our only remaining tent from the summer was our large cook tent, and four catskinners and I were trying to sleep in the -40°C winter weather. I adjusted the oil stove to coax more heat, an adjustment not exactly recommended in the manual. In the middle of the night the stove exploded, and our cook tent was totally engulfed in minutes. We all escaped, but thereafter we had to share a tiny trailer that had been hauled in with the Cat train. It was already overcrowded with its current crew of four. Adding another five made it completely inhospitable. I was not very popular.[28]

A jade boulder, carried by ancient glaciers from its source many kilometres away, dwarfs Stan Leaming. Photographed in Two Mile Creek Valley, close to Kutcho Creek, during a geological survey of the ultramafic belts in the region.

In 1980 the glacial boulders that were strewn across the valley surrounding Provencher Lake had all been tested and the good ones removed. Nephro-Jade then shifted their operations back to Ogden Mountain to the south, and Mohawk continued to mine the *in-situ* jade lenses above the valleys near Provencher. Ogden Mountain at that time was divided into two operations: the Far North Deposit, owned by Howard Lo, and the Japanese-owned Continental Jade property, earlier discovered by Larry Owen and then leased to Joe Bell.

At the Far North operation, Kirk Makepeace developed an exceptional deposit of high-grade jade, which Larry Barr, the original stakeholder, had found in 1974. This single stratum produced in excess of a thousand tons of clean, uniform, dark green jade. One of the large pieces, with Kirk standing astride it, was featured in a *National Geographic* article.[29] Larry had also found a single talus boulder on the property, which was trademarked "Northern Pride" by a California jade marketing company. It weighed about 150 tons, and its centre produced an extremely high-quality jade suitable for jewellery. High-grade, black jade boulders were also found nearby.

The year 1981 was a turning point in many ways. It was the start of a worldwide recession, which in turn resulted in the dumping of hundreds of tons of Cassiar Jade onto the international market. Both Mohawk and Nephro-Jade ceased active mining operations that year, and a gruesome twist of fate saw one of the young crew at Far North lose his leg during the final hour of the mining season. It put an extreme damper on everyone's moods, and no one more so than Kirk. After considerable reflection, he decided to get out of the jade business. His first child had recently been born, and with a family to support, he signed up as a Canadian Tire store manager—from nephrite jade to Rubbermaid™. But as the old saying goes, you can take a miner out of the mine, but you can't take the mine out of the miner. Within a year, Kirk was missing the action. In 1982 he formed his own company, Jade West Resources, and optioned the Far North jade claims from Howard Lo. He also struck a deal with Joe Bell from Jade World that allowed Joe the opportunity to operate Kirk's claims in return for a percentage of the jade removed.

Slowly things began to turn around. By 1985, Joe and Kirk had formed a partnership to mine both the Far North and the Japanese Continental Jade properties. However, an untimely heart attack in January 1986 ended Joe's life, and his surviving partner, Jeannie

MacCulloch, sold the balance of the mining interest to Jade West.

In all businesses, timing is often crucial. So it was to prove for Kirk. The year 1986 brought Fred Ward, one of *National Geographic* magazine's most prolific photo-journalists, to the Canadian jade industry. As part of the magazine's cover story on jade,[30] Fred visited the Kutcho and Ogden Mountain mines. During his extensive research, he was amazed to find so much misunderstanding concerning jade by much of the world. He was extremely critical of the misuse of the phrase "Chinese jade" for jadeite. China and jade are synonymous, but it was only nephrite jade that could claim that historical link.

The *National Geographic* article was the first major acknowledgement of the importance of British Columbia's jade deposits and the part they played in the world industry. Fred's fascination with jade led to numerous magazine articles and two jade books—part of his acclaimed *Gem* series. On speaking tours, Fred used every opportunity to influence the gem trade on the significance of Canadian nephrite jade. The year 1986 also represented a major swing away from declining exports to Taiwan to the expanding new markets of mainland China, where import/export regulations were relaxing. Again, it was Howard Lo who opened this market by convincing factories across China to try real nephrite jade from Canada, instead of pseudo-jades (mostly serpentines) from western China or low-grade jadeite from Burma. Between 1986 and 1989 he arranged exports of nearly 1,000 tons of B.C. jade to China. In historical terms this would rival the entire consumption of nephrite jade in that country for the previous millennium. To date, China remains the largest consumer of Canadian jade, not only in the traditional carving markets of Chinese-style jade art, but also in Taiwanese-owned factories that have moved their operations to mainland China, where labour costs are lower.

In 1987, Kirk purchased all the remaining claims on Ogden Mountain from the Japanese and consolidated the sites as one efficient operation. Jade West continued to mine at Ogden until it acquired Mohawk Oil's Kutcho Creek in 1993, at which time all mining was concentrated on the vast deposits of jade 80 kilometres east of Dease Lake. Not only was it more cost-effective to operate one mine at a time, but world consumption for nephrite had leveled off to a steady 200 tons per year, an amount that the Kutcho deposits could comfortably maintain. According to Kirk, the Ogden Mountain mine remains part of the Jade West group and may reopen when conditions are right.

The Polar Jade discovery

Halfway between the town of Dease Lake and the Kutcho Creek deposits lies an area that had been ignored by jade prospectors for years. Although a likely region for nephrite formation, it lacked the glacial and talus boulders that typified the surrounding jade deposits. A determined prospector named Jerry Scopke staked the mountain above the encouragingly named Serpentine Lake and soon discovered nephrite *in situ*. From the very first excavations it was apparent that this deposit was unique: the exceptionally hard nephrite was confusing. Most of the jade boulders removed had complex swirls of brown or light and dark green, and from the buyer's perspective the product was unpopular, as it was difficult to cut. It also broke in unexpected ways when carved and was not uniform enough for most cutting factories. However, Jerry and his wife, Sue, continued to explore and persevered where others would have left the deposit alone. After half a dozen years of difficult mining, the newly exposed, deeper material was more uniform and showed a highly translucent, emerald- and apple-green jade.

Although the Scopkes' deposit was difficult to cut and had a tendency for irregular colours, Kirk Makepeace recognized it as an important discovery in B.C.'s jade history. It was not a large property, but one he saw as an opportunity, if marketed properly, to shake the long-established perception that Canadian jade was synonymous with an inferior product. Here was an unrivalled source of gem nephrite.

In 1995, with partners Earl Matheson, Bob Dickinson, and mine foreman Tony Ritter, Kirk successfully persuaded Jerry and Sue to sell their mine. The new company, Polar Gemstones, named the product Polar Jade™, not only to separate it from all others, but also to begin a worldwide branding of high-grade nephrite. "We didn't want our jade called *Canadian,* or *B.C.,* jade in world gem markets," Kirk later recalled.

With the trademarked Polar Jade, the company was successful in establishing the material as a unique gemstone. Jade dealers from around the world now insist on Polar Jade for their most important jade jewellery and sculptures.

During the 2001 mining season there was a further discovery at the site that Kirk later described as "the greatest single piece of nephrite in the history of jade production." A 21-ton boulder was removed from the talus, and although larger pieces of jade had been mined before, never had a single boulder of gem-grade material approached such a size. Named "Polar Pride," the boulder was cut into two pieces of 11 and 7 tons (after waste had been discarded), revealing nearly perfect specimens of emerald- and apple-green jade. The company recognized the significance of the find and decided not to sell it through traditional markets for jewellery and small sculptures. "We are searching for a world-class project for this once-in-a-millennium discovery," Kirk announced.

Found right next to Polar Pride, and obviously formed by the same geological processes, a smaller five-ton boulder of equal quality was unearthed that same season. By itself it would have been a major discovery. However, it had the misfortune to be somewhat eclipsed by its 21-ton neighbour. Nicknamed "The Emperor's Stone," it was considered too perfect to be cut up and instead was treated as a (very large) mineral specimen. Parts of its surface rind were ground away to reveal the nearly perfect interior, and it was later shipped to the Munich 2004 Gem Show, where it was sold with the tag line "finest and largest gem grade jade ever seen in Europe."

Cassiar Asbestos Mine

A long time before the coming of Europeans, the Tahltan people noticed that the local mountain goats often had yellow wool clinging to their coats and that birds' nests made with the same fibre did not burn when thrown in a fire. What they had found was a high-grade form of asbestos. However, it was only after the Alaska Highway project of the Second World War had improved access in the north that a serious look was taken at this curiosity.

In 1949 a party from the Geological Survey of Canada spent four days prospecting McDame Mountain in the northwest corner of the province, but they concluded that the area was too remote for any near-future development. Nevertheless, just a year later, in June, Vic Sittler set out with two pack dogs, some provisions, and a rifle, and after climbing 1,900-metre-high McDame Mountain, he marked out two claims, took 10 kilograms of sample fibre and headed for home. En route his dogs held a grizzly at bay until Sittler could shoot it. The remainder of the trail was all the slower because of the additional bear hide.

Vic Sittler had staked the centre of a lode that featured exceptionally long chrysotile fibres, excellent for spinning heat-resistant cloth and for use in other industrial applications, such as reinforced concrete and as the base of automobile brakes and clutches. Commonly called asbestos, it is a hydrous magnesium silicate. A pilot mill was built in 1952, and the Cassiar Asbestos Mine was born. Initially production was trucked north via the Alaska Highway to Whitehorse and then down the White Pass Railway to Skagway in Alaska, but this was clearly an expensive route. The opening of the highway south to Stewart allowed better access to an ocean port (Prince Rupert), and thus was the impetus for the highway's expansion.

At the site, the company consigned all the rock overburden to the dump as unrecognizable waste. One of the company's mechanics, Clancy Hubble, was a rockhound, and in 1967 he found a strange rock on the property's #6440 bench. He had it identified by the B.C. Department of Mines in Vancouver as nephrite and Clancy's interest was triggered. Alf Wanke, another mine employee, is also reported to have found a rock that was later identified as jade. Three years later Clancy spotted a boulder on the slide at the waste dump and realized there might be as much as a thousand tons of nephrite jade lying in a dusty heap. He talked the mine's management into saving a few tons for him and got them to agree to his staking the dump, which lay above part of the mine's holdings. At that time, the company regarded the material as a nuisance, as it was extremely hard to drill. It appeared in small pods along the hanging wall alteration zone between the serpentinite and the overlying argillite and was treated like the other waste. The removal crew got to the jade pods about once every two to three years. Now, after the claim was staked and the bench with the hard green rock had been blasted, the jade was separated during loading and then stockpiled.

In return for the claim, Clancy reached an agreement with the company by which they had a share in his jade-carving business. It no doubt came as a surprise to the managers that the mine's "waste" was, in fact, worth more by the ton than their high-grade asbestos product. However, it quickly became obvious that Clancy could not process a fraction of the jade being cleared from the pit. In his spare time he made jewellery that he sold to local residents who wanted unique gift items.

Even when the mine began to sell the jade commercially it was still a small-time affair, and most of the proceeds were turned over to the town. The Cassiar Kinsmen's Club built a community facility that included a swimming pool and ice rink. But gradually, the strong demand from Taiwan raised the company's awareness of this resource, which accumulated at 100 to 200 tons annually.

The mine management approached Don and Gwen Lee to market their product to overseas buyers. However, they declined, as Don was at that time committed to sell Cry Lake's jade and felt that if he represented the two main players, there might be a conflict of interest. He suggested that the Cassiar management talk to Cry Lake, and if they could agree on terms, the Lees would then represent both. This in fact did happen, and thereafter, the Lees' Langley yard looked even fuller at times.

All the jade was counted onto the plane at Cry Lake/Kutcho and then again onto the truck bound south from either Dease Lake or Cassiar, so there was a means to track and account for the boulders. But one boulder often looks like another boulder. Was there ever uncertainty about which material came from which site? Not according to Don Lee. For a start, the Cry Lake material was flown out in a Twin Otter (then later in a Caribou), so it never weighed more than 1,500 pounds and was usually sawn.

The Cassiar material, by contrast, was usually in massive chunks, given that the mine had large earth-moving equipment. Early on, the jade showed signs of blasting, but as the miners realized what a resource it was, they became more careful at removing it from the pit, and so it was clean of stress cracks. The Cassiar material also had characteristic red paint identifying it at the site, so when it was all graded and laid out in Don Lee's yard, the two sources were fairly easy to separate visually. Finally, the Cassiar nephrite had a green fleck that was known as "chrome jade." Initially the Oriental buyers didn't want it, but later preferred it and paid a premium to get a supply.

Most of the mine's nephrite production was sold to Taiwan. It was reported that a 320-ton sale in 1981 brought in $530,000 (or about 83 cents a pound), and was worth a $30,000 commission to Cry Lake—a tidy amount in those days.

Clancy Hubble moved to the Prince George area in 1974, where he continued to produce jade artifacts from the material he had acquired during his years with the mine. Cassiar Asbestos closed in 1992, after 39 years of producing an exceptionally high-quality form of asbestos completely free of tremolite. It is tremolite's sharp-edged mineral crystals, found in much of the world's production, which are considered so hazardous, as they cause asbestosis, a fatal lung disease.

Some effort is being made to recover the jade that was consigned long ago to the tailings piles as waste. Jedway Enterprises recovered 50 tons from the old dumps in 1998. The tailings are also a potential source of the metal magnesium, and a 2002 report suggests there are about 3.6 million tons of magnesium on site. In 2000 Cassiar Asbestos reopened as Cassiar Magnesium (serpentinite has about a 26 percent magnesium content). There was talk that the mine might convert from open pit to underground, and if that does happen, the jade will be lost, as the overburden would no longer have to be removed to reach the ore.

~ Yukon ~

The geological conditions in the Yukon are an extension of those in the B.C. tectonic setting. The serpentine belt in the middle of the province continues north across the provincial boundary, so it is hardly surprising that jade has been found there. As early as 1888, the GSC geologist G.M. Dawson noted there was jade float in the upper part of the Lewes River (a tributary of the Yukon). In his field report Dawson noted:

> Having become interested in the origin of nephrite or jade, on account of its former intensive employment by the Natives of the West Coast for the manufacture of implements, I kept a close watch for this mineral along our route, and ultimately succeeded in finding several rolled pieces of it in gravel bars along the Lewes. Of the pebbles collected by us, at least five have the specific gravity and other physical characteristics of jade, though they have not been subjected to chemical or microscopical analysis. Several of these are evidently, however, pure and typical jade, of which the finest and most characteristic was found by Mr. W. Ogilvie, near Miles Canyon. This specimen is a pale green translucent to sub-transparent variety, weighing a pound and three-quarters … So far as I have been able to ascertain, the discovery of jade here noted is, with one exception, the first actually direct one made in the region of the Pacific slope. The exception above alluded to is that of jade found in the Kwichpak mouth of the Yukon during Captain Jacobson's stay in that vicinity and which was obtained by him and taken to Berlin.[31]

In 1953 a prospector reported the presence of an apple-green stone, thought to be jade, from the slopes of the mountain east of Klukshu Lake, close to the Haines Road, and about 40 kilometres south of Haines Junction. No serpentinite was noted in the area, and despite a two-day search, no source was discovered for the stone. Outcrops are covered by deep glacial overburden, but there were a reasonable number of exposures to warrant a careful search. There are specimens in the Metropolitan Museum in New York taken from the south end of Klukshu Lake at an elevation of 1,200 metres.

A contact reaction zone of the rodingite type was seen in serpentinite, west of Lake Laberge, near Whitehorse. The direction of prior glaciation suggested that ultramafic bodies on Jubilee Mountain may have contributed to the alluvial nephrite found near Miles Canyon. Elsewhere in the territory, an asbestos and jade prospector by the name of Roy Sowden found *in situ* deposits of jade in the mountains west of Mile 84 on the Robert Campbell Highway (Highway 4) in the Yukon, close to the B.C. border, in the 1970s.[32] The deposit was near Frances Lake, and Sowden staked a number of claims. Although the *in situ* lenses and boulders were relatively close to the highway, they were at a high elevation, and with the remoteness of the locality and the cost of transportation, any future jade mining was effectively ruled out. At one point, though, a few tons were removed by helicopter. Attempts were made to put in a road, but they were thwarted by permafrost, which turned the track into a quagmire during the summer. However, in 1978 Karl Ebner hauled out a 31.5-ton boulder that made it into the *Guinness Book of Records*. At that time, probable reserves were estimated at between 300 and 1,000 tons, but mostly of low quality.

Jade does not always form in massive lenses at contact zones. Very occasionally it develops in grapelike clusters, known in geological terms as *botryoidal* jade (from the Greek word *botryoides*, meaning grapes). This form was first identified in the 1950s at Jade Cove in California.[33] A few other sites have been identified in Mendocino County, California; Cultus Mountain, Washington; Blue Moon Jade, Arizona; and the curiously named town of Happy Camp, California.

During one Yukon summer field program, Roy Sowden saw what he took to be a bird's nest under a ledge across a creek when he stopped for a break. Closer inspection revealed that it was an extrusion of botryoidal jade. It was barely a metre long and wide, but with the aid of his portable saw he was able to free a sample.

Prospectors are busy people, and some things get left on the back burner. The botryoidal jade from Frances Lake was just such an item, and over two decades passed before the specimen came into the possession of Duke McIsaac, who spent considerable time releasing the jade nodules

Stan Leaming (right) and Karl Ebner with a cut block from Karl's jade property near Frances Lake, just north of the B.C.–Yukon border. An unpredictably short summer season and the high cost of extraction and transport has to date prevented the full potential of this sizeable nephrite outcrop from being achieved.

from the host matrix, after which they made striking collector's specimens. A single lens of nephrite jade weighing 577 tons was found in the Yukon by Max Rosequist in July 1992. It is owned by Yukon Jade Ltd.

Jade Craft Arrives in North America

The Stone Age was the first cultural name in the classification of human advancement, to be followed in turn by the Bronze and Iron ages.

The Stone Age itself is divided into the Paleolithic Period (when tools were formed by chipping stone to make artifacts such as arrowheads, knives, adzes, etc.) and then, with advancing technology, the Neolithic, when items were manufactured by grinding away, rather than chipping, extraneous material.

These Stone Age divisions were not time periods. Rather, they were cultural stages, which varied from place to place. In British Columbia, some cultures were in the Neolithic Period until 1750 AD, whereas in China the oldest Neolithic jade artifact dates back 6,500 years. The period lasted until the beginning of the Bronze Age in about 1000 BC.

Bronze is produced by fusing together copper and tin. The molten product can be cast into any shape in a mould, and some very large artifacts can be produced. It was a major cultural achievement, and when various societies manufactured bronze, stone tools were quickly relegated to refuse dumps (to the delight of archaeologists in succeeding centuries). Some cultures entirely abandoned the use of stone for tools and weapons. Others kept select items to fashion into ritual or ceremonial objects. This was particularly true in China. Throughout the many ensuing dynasties, nephrite was treasured and carved into a wide variety of objects that had particular significance for the owners.

Similarly, for thousands of years the indigenous peoples of Siberia processed nephrite into tools, weapons, and ornaments, using deposits in the East Sayan Mountains and elsewhere. Radiocarbon dating from Siberia reveals an ancient jade history there. In 1000 BC, not only adzes but also white nephrite rings were produced. The latter are thought to have had some ritual significance. Some of the raw material found its way via trade routes to the workshops of the emperors of ancient China.

The history of jade in China goes back at least 6,000 years, and some researchers have postulated that a Jade Age preceded the Bronze Age. Presumably many of the deposits are now depleted, for it is a depreciating asset, and jade outcrops are generally small. Jade was certainly found in Chinese Turkestan, now called Xinjiang Province, from where it was transported to eastern China. Possibly jade from the vicinity of Lake Baikal in Siberia made its way to the Chinese artisans as well.

In regard to still earlier times, it is not yet known with certainty the origin of the many groups who made their way to North America via the Bering Strait during the time it was a dry land corridor. Perhaps many of the tribes wandered from Asia to the northeast and brought their jade tools and weapons, plus the methods of making them. There was plenty of room to move into the uninhabited land between Siberia and Alaska (called Beringia by paleogeographers). Those reaching the Kobuk area of Alaska found ready sources of jade from *in situ* deposits, and alluvial boulders and cobbles were plentiful 250 kilometres from the mouth of the Kobuk River, where several tributary streams contain river-borne nephrite. These boulders were derived from lodes developed in the serpentinite to the north, in what would later be called the Cosmos and Jade mountains.

Great herds of animals inhabited the plains of Beringia, which included much of what is now northeast Siberia and northwest Alaska. There were mammoths, bears, camels, and many edible ruminants, so we can assume that the natives of Asia followed the herds until the North American continent was reached. Current archaeological evidence suggests that these people were the ancestors of the present-day American Indians.

Later, other groups more adapted to the Arctic environment followed and remained in the North. These were the ancestors of today's Inuit (formerly called Eskimos—"eaters of raw meat"). Other migrants came to inhabit Alaska's Aleutian Islands. Of different origin, the Aleuts were an ethnic group who had diverged from a common stock with the Inuit and had moved into southern Alaska some 8,500 years ago.

Anthropologists still debate the time and mode of the first arrivals. The case has been made for immigration from Africa or Asia via oceanic passage, and this finds varied favour with professional anthropologists. University of Kentucky archaeologist Dr. Tom Dillehay threw out a challenge when he unearthed a 12,500-year-old settlement in southern Chile, implying that the first migrations across Beringia must have been at least 15,000 years ago, or even earlier, which seems unlikely given the presence of extensive glaciation then.

One may also wonder why the inhabitants of South and Central America were able to reach so high a civilization, as did the Olmec, Maya, Aztec, and Incas, if they had not been there much longer than their neighbours to the north. These civilizations had sophisticated oral languages and calendars and knew something of astronomy. They carved both architectural and ornamental stone to a very high level of sophistication. Metallurgy was widespread, with copper and gold being smelted and worked. These accomplishments indicate a lengthy time for development.

Some propose that the migrants arrived in small family groups over an extended time, beginning as long ago as 30,000 years (the "trickle theory"). Others suggest that there were mass migrations on a few occasions (the "wave theory"). It is generally agreed that some, if not all, migrants did come from northeastern Siberia via the land bridge now flooded by the waters of Bering Strait or, alternatively, by boat, since the distance was not great at that time.

The amount of water stored in the extensive ice caps in the Pleistocene epoch two million years ago (MYa) was sufficient that the sea level was lowered a hundred metres—enough to expose the sea floor and allow the transfer of plants, animals, and humans at that time. At first, the way south would have been blocked by ice sheets grinding down out of the Cordillera into what is now Alaska, British Columbia, and Washington on the west, and Yukon, Alberta, and Montana on the east. Probably many groups remained in the Alaskan ice-free zones, where food would have been plentiful.

Current thinking suggests that as the ice melted, migrants made their way south along the shoreline or via the interior. The great ice sheet that had spread from the Cordillera into the Prairies, where it merged with the

Canadian Shield ice sheet in the east, had begun to recede, forming an ice-free corridor down the eastern side of the Rocky Mountains. Through this passageway, many groups of people spread out into the Great Plains, moving steadily south. Whatever the method of migration, recent archaeological findings suggest that when the ice retreated about 11,000 years ago, there were established communities, fully adapted to their environments, in such diverse localities as the plains of the Midwest and Argentina, the jungles of tropical Brazil and Columbia, the arid *altiplano* of Bolivia and Peru, and the frigid ice deserts of the High Arctic.

Early European explorers noted the importance of jade to a variety of cultures, both to the north and south of what is today the United States. In Central America, for example, artifacts revealed that jade was the single most important item of value in a Paleolithic society that put a high value on various minerals.

> Many specimens of carved jade were brought over early to Spain ... Wonderful tales were told of the sacred articles of 'emerald' belonging to Montezuma, including a goblet and 'rose' that were shipped by Cortes to the King of Spain; amongst the choicest treasures of the Conquest ... It is impossible that they could really have been emerald, as that stone scarcely occurs in Mexico at all. They were probably *chalchihuitls* of peculiar richness of colour, and constituting doubtless both in workmanship and material the finest products of Aztec art.[1]

~ Jade in North America ~

To the north, in 1778, Captain Cook noted the similarity between the axes and chisels used by Natives along British Columbia's west coast and those of the New Zealand Maoris, whom he had visited shortly before. In 1786 the French navigator La Parouse visited Alaska and recorded jade implements being used by the Natives on the west coast of Prince of Wales Island (off the Alaskan Panhandle) as being "so hard as to cleave the closest wood without turning its edge ... "

While "jade," "emerald," "jasper," and even "turquoise" appear to have been used interchangeably by early Europeans, it is usually self-evident when and where each is actually being described. The jade of North America north of the Mexico–U.S. border is the nephrite variety, except for a few occurrences in California and one in Washington State. Certainly all the jade from archaeological sites is nephrite.

Jade is found in North America principally in two belts of rocks, along the east and west margins of the continent. The eastern, or Appalachian, belt is of only minor importance. There are known sites in Newfoundland

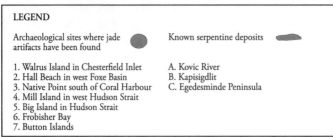

Map 4: *Neolithic jade artifacts have been found in the eastern Arctic around Hudson Bay, although the source of the material is unclear. Known deposits of serpentinite are shown.*

and possible occurrences in Quebec, Vermont, and North Carolina. The western, or Cordilleran, belt extends from Alaska to California. There are a number of terranes that originated in the Pacific Ocean and have, over many millions of years, accreted onto the western margin of the continent. Jade occurrences are virtually unknown in the centre of the continent. There are occasional references to artifacts, tools, and cobbles, but these are likely the result of Neolithic trading.

In the Canadian Arctic, numerous jade artifacts have been found in Inuit sites or in those of their forebears. The numerous discoveries around the Fox Basin suggest some local source, and there is a reported site on the Rae River that has proven elusive to later prospectors. In northern Labrador there are three belts of ultramafic rocks in the Cape Smith fold belt, which crosses the north end of the Ungava Peninsula. And while serpentines do occur in Greenland, and many jade artifacts have been found there, jade deposits are unknown to date.

With the advent of European culture in the sixteenth century, the use of jade declined rapidly, as iron tools replaced jade implements. The Cordilleran jade fields were deserted until the California gold rush of 1849. A decade later, when the Fraser River gold rush started in British Columbia, the Chinese labourers were the first to identify the "stone of heaven." There is evidence they shipped cobbles back to China in the ensuing years.

With the decline of the small-time prospector, the jade fields were forgotten again until after the Second World War, when a rise of interest in lapidary art and rockhounding thrust jade once again onto the North American stage. Beginning first in Wyoming and California, and then later in B.C., the jade fields assumed great importance, and with ready buyers in Germany and the Orient, the industry grew rapidly. Although the unit value of jade did not match that of gold or other mineral commodities, it nevertheless attracted interest, and numerous new sources were found, including *in situ* lodes. The commercialization of jade became an important economic benefit. Records of production were kept by the B.C. Ministry of Mines and Petroleum Resources starting in 1962, and in 1978 the annual value of British Columbia jade sold topped a million dollars for the first time.

~ Jade in British Columbia ~

Sometime between 5000 and 10000 BP, the Native newcomers to the continent made their way down the coast of the Queen Charlotte Islands and into the archipelagoes surrounding Vancouver Island. There may have been rapid migration by way of a maritime route from the north by tribes who were able to construct dugout canoes, such as those of the Haida Gwaii in historic times.

Jade was part of the Inuit culture, and both *in situ* and alluvial boulders were available from Alaska, which was within trading distance of the western Arctic. It is conceivable that this material had been traded to the south and had come in contact with the trade from the Fraser River. It is known that nephrite among the Tlingit tribes (resident along the Alaskan Panhandle) was attributed to trade from the south, despite being close to northern sources.

The Tlingit regarded jade (which they called *tsu*, meaning "green") with respect. It is worth considering that some nephrite may have been derived from the Kobuk River in northwest Alaska. Research suggests a possible source in Tlingit territory, near the head of the Duncan Canal. Certainly, the Tlingit prized the material highly, for Lieutenant George Emmons says that a jade adze blade was worth "from one to three slaves."[2]

"According to the Tlingit," reported Emmons, "so important was a jade adze that when its owner used it, his wife was expected to refrain from frivolity. Unbecoming conduct on her part might result in the jade breaking." Since jade is universally tough, it would appear that the Tlingit women were given more credit for quiet behaviour than they were due.

The Haida, too, had nephrite tools, possibly traded from the south or captured in war party expeditions, for they were great seamen and ranged widely. Other nations along the coast were also in possession of nephrite artifacts. The Nuu-chah-nulth on Vancouver Island likely acquired theirs from the Lillooet area through trade, as they were in contact with the tribes of the Fraser Valley and delta.

Lieutenant Emmons travelled extensively in B.C. and Alaska in the early twentieth century and amassed a large collection of jade artifacts, which he later exhibited.[3] However, his travel notes and observations were even more interesting. At the mouth of the Fraser River and on the east coast of Vancouver Island he unearthed jade celts (chisels) in considerable numbers. He believed they came from centuries of trade and migration.

Emmons also found old camp and village sites, where quantities of partly worked boulders, as well as cutting tools and finished implements, were found "in the possession of the older people, which have descended to them from the past. Jade in both the rough and finished state was the most valuable article of trade possessed by the natives." Elsewhere, he bought "not without difficulty several small, beautiful deep green pebbles, which were kept with personal belongings. Such are said to be worn suspended as ornaments by girls and women of earlier days." Among the B.C. coast's Salish people, jade was known as *la'ist*, meaning "green stone."

There were also jade deposits in the territory of the Kaska and Tahltan people in the vicinity of Dease Lake, as well as in the territory of the Sekani and Carrier (now known as Dakelh) Natives in the Fort St. James area. However, none of these people worked those deposits, as far as is known. A few artifacts have been recovered in the northern regions of the province, but it is assumed that they were obtained in trade from the Lillooet area.

Farther south in Washington State, some of the tribes had access to jade from sources in the vicinity of Darrington. The Stillaguamish River and its tributaries contain alluvial jade, and some of this material probably found its way to the coast and the San Juan Islands (Orcas, Vashon, Whidbey), where artifacts as well as boulders have been found. Nephrite cobbles at the mouth of the Nooksack River, just south of the Canada–U.S. border, have also been reported.[4] Nephrite celts have been found in the vicinity of Marietta, Washington, so there seems to have been a local industry using a local source, possibly not requiring trade from the Fraser River sites. Jade from that source, however, did find its way into the northern part of Washington State via the Okanagan Valley.

Farther south in Oregon and California, nephrite has been found in boulders and *in situ* deposits, but it seems that early Native use of the jade there was relatively insignificant—although a few artifacts of both nephrite and jadeite have been reported in the literature. However, it does raise the question of why Stone Age societies in the north (Alaska, Yukon, and B.C.) prized and used nephrite extensively, as did those in Central America (who used the very similar jadeite), while the societies between them did not. Perhaps it was the differences in lifestyle.

It is worth noting that at the time of European contact the tribes of the north were relatively settled and were moving away from a traditional hunter-gatherer culture. Even more so, the societies of Mesoamerica had reached a high level of agricultural stability. Both these lifestyles involved a settled or seasonally settled existence, stable food supplies, and hierarchical societies that allowed for chiefs, priests, and an educated class. All of these developments led to an environment in which arts and crafts could develop. The tribes between these two regions, however, still retained their traditional hunter-gatherer existence. Even when exposed to new concepts like food growing, domesticated animals, and advanced crafts, their transient lifestyles simply prevented them from adopting anything more than the simplest items.

For nearly 4,000 years the ancestors of the Lillooet people living near the confluence of the Bridge and Fraser rivers found and used alluvial jade cobbles for the manufacture of tools and weapons. Boulders were of no use to them, as they were simply too large to manage. But the inhabitants were blessed with a supply of small pieces that was more than enough for their own needs. Trade for the valuable excess material extended to the coast and into the interior. However, jade was a difficult medium to master. It required the abrasion of natural substances like thin-bedded sandstone or schist, which were only a little harder than the jade itself. With much time and patience it was possible to make thin, parallel-sided blanks from the river cobbles, and then, with more patience and time, an edge could be ground on one end to make an efficient cutting tool. A few of these artifacts have been recovered and are in museums and private collections.

A century was to pass before Dr. George Dawson, who later became the director of the Geological Survey of Canada, reported in 1887 on the occurrence and use of jade by the Natives of British Columbia.[5] At that time, there was no known source for the alluvial deposits. Dawson speculated that the origin must be in the "highly altered and recomposed volcanic rocks of the Carboniferous and Triassic." He also observed that the method of working jade was with " … a thong or thin piece of wood, in conjunction with sharp sand." He noted that adzes were the usual tools manufactured.

There were other sources of nephrite in the interior of the province, but few had the advantage of the power of the mighty Fraser River in reducing large boulders to manageable sizes. Some artifacts have been found in the northern part of the province in the vicinity of Dease Lake, where jade *in situ* and alluvial deposits are known and where one might expect some Native use to have developed. But we do not know if the objects were traded items or locally manufactured.

The Science of Jade

 C onsider rocks. They are the material of older buildings, of tombstones, and a medium for artistic expression. Look at any rock or stone closely, and you will see that it is made up of many parts. A few rocks may be essentially pure, in that only one component is present, and that component is called a mineral. In the case of a marble statue, for example, you are looking at the mineral *calcite* (*calcium carbonate* in technical terms). A granite tombstone, on the other hand, will have two, three, and even more minerals. The whitish material in granite is almost certainly quartz, (*silicon dioxide*), the pinkish component is a *feldspar*, and commonly the black mineral may be *mica* or *hornblende*. All are silicates: that is, they are built up of silica and oxygen, with other elements attached.

Jade belongs in the silica category, but before getting technical, jade can be defined in a non-technical way. First, jade is not a unique substance. The name is a generic one, as previously mentioned, signifying one of two separate and distinct mineral aggregates. One is called *jadeite*, the other *nephrite*.

Jadeite is an unusual member of a common family of minerals called the *pyroxenes*. It comes principally from Burma (now Myanmar). It is found in a few other localities, but rarely with the superior lapidary qualities of the Burmese material. Because of centuries-old antagonism between Burma and China, jadeite was only really accepted in China during the reign of Emperor Ch'ien-lung (1736–1795). The Empress Dowager Tz'u His took a liking to its bright colours, which can vary through white, cream, grey, smoky blue, lavender, yellow, orange, brown, and green to black.

Nephrite, on the other hand, belongs to the *amphibole* group of minerals. During China's Stone Age some four to five thousand years ago, royalty prized a hard, tough rock, commonly green in colour. Nephrite is not necessarily green, however, and the Chinese in fact preferred an off-white version, referred to as "mutton-fat jade." Nor does the English word "jade" have quite the same significance as their word *yu*, which means the more general term "precious stone," because more than one mineral qualifies for that title. During the late Neolithic Period (3000–1500 BC), nephrite jade began to be imported from the western regions of China. There seems to be no record of *in situ* deposits in China itself, but boulders were found in the rivers of East Turkestan, and these found their way to the cutting houses on the Chinese plains.

~ *Jadeite* ~

Jadeite is a mineral species established by the geochemist A. Damour in 1863, and it differs markedly from nephrite in that its relation lies with the pyroxenes rather than with the amphiboles. Chemically it is an aluminum sodium silicate, $NaAl(SiO_3)_5$, related to spodumene. Its colour is commonly very pale, and white jadeite, which is the purest variety, is known as "camphor jade." In many cases the mineral shows bright patches of apple- or emerald-green because of the presence of chromium. Jadeite is much more fusible than nephrite (jadeite contains water molecules in its matrix, which boil off as steam when a sample is heated), and is somewhat harder (6.5 to 7). However, its most readily determined difference is its higher specific gravity, which ranges from 3.20 to 3.41. Some jadeite seems to be a metamorphosed igneous rock.

Burmese jade, discovered by a Yunnan trader in the thirteenth century, is mostly jadeite, as previously mentioned. The quarries are situated on the Tjiu River about 200 kilometres from the village of Mogaung in Upper Burma, where jadeite occurs in serpentine and when quarried is partly extracted by firesetting, a method used to heat and crack large boulders. It is also found as boulders in muddy sediments (alluvium). When these boulders occur in a bed of iron-rich mud (laterite), they acquire a red colour that gives them additional value.

According to early visitors to the jade country of Upper Burma, jadeite occurs at three localities in the Kachin Hills: Tawmaw, Hweka, and Mamon. It is known locally as *chauk-sen* and is sent either overland to China or down to Mandalay on the coast, by way of the town of Bhamo. As a result, Bhamo has erroneously come to be regarded as a locality for jade.

Jadeite also occurs in association with nephrite (see below) in Turkestan in Western China and possibly in some other Asian localities. In certain rare cases, nephrite is formed by the alteration of jadeite. The Chinese *feits'ui*, sometimes called "imperial jade," is a beautiful green stone, which seems generally to be jadeite, but it is said that in some cases it may be chrysoprase, which is a green quartz. The finest grades of Burmese imperial jade approach emerald in value and colour. They have recently become a favourite of Asian people and have given rise to the misconception that jadeite is "Chinese jade" and superior to "Canadian jade." In fact, the jade of the ancient Chinese dynasties was actually nephrite, the same mineral found in British Columbia, in the western states of the U.S., Siberia, New Zealand, and Australia.

Jadeite is also the "jade" of Central America. Recent discoveries (or more correctly, rediscoveries) of ancient Mayan jadeite quarries in Guatemala have greatly increased our understanding of the development of jade cultures in that region. There are other Mesoamerican jadeite deposits as well, notably in southern California. Elsewhere in the world, a high-grade jadeite (comparable to the gem quality of Burmese Imperial Jade) was discovered in the early 1970s in the northern Ural Mountains of what was then the USSR and is now Russia.[1] However, the extreme remoteness of the area, the poor transport infrastructure, and the difficulty of securing foreign capital have all combined to prevent the prospect from developing into anything more than a curiosity at this time.

~ Nephrite ~

The second type of jade is *nephrite*. It is actually the common mineral *tremolite*, but with a difference: the habit of the mineral is fibrous. The fibres are intertwined and matted into a dense mass of bundles in a random orientation. Just as rope gains its strength from a myriad of tiny fibres, so too does nephrite get its great toughness, which allows it to be carved into delicate forms. This means that a sharp edge, once formed on a tool such as a knife or an axe, will keep that edge almost indefinitely. Research published in 1973 showed that both nephrite and jadeite have higher compressive and tensile strengths than any other rock aggregate or industrial ceramic.[2]

There may be other minerals within nephrite, such as green splashes of chrome garnet or black spots of magnetite, which mar the appearance and downgrade the material for lapidary purposes. Nephrite is commonly some shade of green but may also be nearly white or very dark green to black in colour. Very rarely, it may take on a bluish or brownish shade.

~ Etymology: What's in a name? ~

The actual word "jade" comes to us from the Spanish phrase *piedra de yjada*, or "stone of the loins/kidneys." In their conquest of Central America, the Spaniards were interested only in gold, but the Aztecs prized one stone above gold, much to the amusement and amazement of the *conquistadores*, who called it "green jasper" or "emerald."

Latinized to *lapis nephriticus*, or "kidney stone," its medicinal properties were thought to be effective for diseases of that organ, and the belief was brought back to Europe in the early sixteenth century. Medical science was then in its infancy, while diseases were rampant, especially in the tropics. Frightened people believed readily in a wide array of talismans, herb concoctions, and rituals. In due course, the *yjada* of the Spaniards became *l'ejade* of the French, then later *le jade* in Brittany, and finally *jade* in English. The word, of course referred to the specific mineral from Central America.

While the Spanish were conquering the New World and sending back new and unexpected discoveries from the Americas, such as potatoes, rubber, squash, and corn, those other great navigators, the Portuguese, were rounding Africa, bound for India and China. In 1557, in Macao, they learned that the Chinese prized a hard, sometimes green, stone that was called *yu*. The Jesuit monks, who were particularly conversant with alternative medicine, thought it closely resembled the *piedra de l'ejada* of the Spanish. The two languages being similar, the Portuguese called the stone *piedra de mijada*.

In the sixteenth century Latin was the language of scholars. "Jade" was translated into the world of science as *lapis nephriticus* (literally, "stone of the kidneys"), and from that universally accepted nomenclature we get today's common name *nephrite*. However, it was only in the late twentieth century that the mineralogical structure of the stone from the Mayas and the stone from China were shown to be quite different. Indeed, from a scientific point of view, the two minerals (or aggregates, as they should more correctly be termed) should not have been given the same name. However, the name "jade" has stuck. Today "jadeite" is the jade of Central America (Guatemala and Mexico), Myanmar (Burma), and a few other places of minor importance. "Nephrite" is the jade of British Columbia, China, Siberia, New Zealand, and several other places too.

~ Recently named "jades" ~

Amateur mineral collectors and creative promoters have come to realize that the word "jade" has an exotic connotation, which can enhance what might otherwise be a lacklustre, non-jade mineral. As a result there are a bewildering number of "jades" to be found in rock shops, gift stores, and even at high-end jewellery stores. Let the buyer beware! Most have no connection with jade or nephrite at all. Following are some of the more common trade names and their actual definitions.

- Amazon jade and Colorado jade are usually amazonite (a green feldspar).
- American jade is usually Californite (a green variety of idocrase).
- Imperial Mexican jade and Mexican jade are usually green-dyed calcite.
- Indian jade is often aventurine (a green quartz).
- Korea jade and new jade are usually serpentine (a magnesium silicate).
- Transvaal jade is really green (grossular) garnet.

Jade in North America and Europe

 The western jade belt includes all of the mountainous area from Alaska through Yukon Territory, British Columbia, Washington State, Oregon, and California. Mexican jade, while nominally the same structure, is primarily jadeite and is not dealt with here. The western continental margin is recognized as being a collage of exotic terranes that originated out in the Pacific Ocean and that have been "smeared" against the westward-moving North America plate over a period of many millions of years. Some of these terranes merged before hitting the continent, while others were single collisions. Some were built on an ocean crust, with a serpentine basement on which volcanics, cherts, and limestone accumulated. It is the serpentines that gave rise to the Cordilleran jade deposits. The Wyoming deposits are an exception.

The docking of these terranes was not a passive event. Great forces resulted in deformation, faulting, and associated metamorphism and vulcanism. Nor did it occur head-to-head. Most collisions were oblique, resulting in strike-slip faults—a process that occurs to this day in such locales as the San Andreas Fault in California and the Wrangellia Plate off Vancouver Island. As a result some terranes are stretched along great lengths of the continental margin. Geologists now recognize over 200 terranes in western North America, although it should not be supposed that these all represent individual terranes moving in from the Pacific

Ocean. Some are the torn remnants of larger terranes and may have developed after their collisions.

~ Alaska: The mystery ~ surrounding Jade Mountain ~

The story of jade in Alaska is inextricably tied to the Aboriginal peoples of the area. Early explorers noted the use of jade implements—especially axes, picks, and celts—by the Salish people of lower B.C., the Tlingit of lower Alaska, and the Inuit of the north.[1] The origin of this jade was hotly debated until 1883, when a report by E.W. Nelson revealed that in his travels from Norton Sound on the Bering Strait all the way to Point Barrow in the north (on both Alaskan and Siberian coasts), the owners of all the jade tools named "Jade Mountain" in Alaska as the source of the jade. Nelson wanted

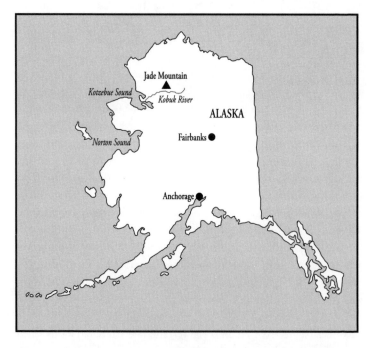

Map 5: *The extreme remoteness and short summer season meant that tracing the source of jade in Alaska was a formidable challenge for early explorers.*

to find this mythical peak, but a combination of hostile Natives and short summer seasons prevented him. However, he noted:

> These people all agreed in the statement that the jade occurs on the side of a steep hill or mountain slope descending into one of the rivers, and each described its occurrence only along what appeared from their descriptions to be a well-marked vein, or perhaps a dyke, extending from the water to the crest of the hill.[2]

Naturally, such a report triggered considerable exploration and even more speculation, since it had an almost *El Dorado* mystique to it. Credit for solving the riddle went to the aptly named Lieutenant George M. Stoney of the U.S. Navy, who tried three times to reach the peak. He reasoned that access to the mountain was best achieved by using a boat because of the difficult coastal terrain, and he knew that as early as 1850, Admiral Beecher of the British Navy had described a river that allowed access to the region from Hotham Inlet in Alaska. The key to the mountain was to find the river. Of his second attempt, Lieutenant Stoney later recalled:

> The native name of the river is Kowak [Kobuk]. I noticed the natives had many implements of jade stone, and several pieces of stone in the rough state. I inquired through an interpreter where they had gotten it, and was told 'on the big river', meaning the Kowak River for which I was then searching … As we passed on, the mountain was pointed out by the natives 10 or 15 miles off to the north, and was seen by me. When 110 or 115 miles from the mouth of the river, and opposite or nearest to the mountain, I landed, proposing to march over the tundra to visit it. The natives refused to accompany me, owing to superstitious fear. A native pointed out the best way. I took a white man and started, but was told again and again we would not return.[3]

They did return, but with green stone samples that subsequently proved to be serpentine, not jade. Undaunted, George Stoney set out on his third expedition, after over-wintering in Fort Cosmos.

> In July 1886, I undertook the exploration of 'Jade Mountain' going with the steam launch to the spot nearest the mountain, which happened to be the site of a deserted Eskimo village of a few huts … then I started overland. The expedition consisted of myself, two white men, and five natives, with four days' rations. The walking over the tundra

was severe, and the mosquitoes terrible, both by day and by night …
From a distance (Jade Mountain) looked as though covered in grass.
The natives have a legend about this mountain: that it was the first dry
spot on earth: that the world was covered with water, when a crow flew
over and stuck a long staff in the ground. The water then commenced
to recede, and the peak of this mountain appeared. After plenty of
dry land was seen the crow turned into a man … I noticed that the
exposed rock was softer than that not exposed. The harder portions,
which after proved to be jade, came in strata like quartz, the strata
varying in thickness … The jade gotten out by the natives for their use
is said to have been done by their shamans or medicine men …

Being a naval man, Stoney fixed the mountain's position with nautical
precision: "Jade Mountain (lat. 67^0 66'N. long. 158^0 04'W) … The height
of the mountain is about 1,000 or 1,500 feet above the river valley." Others
who came later commented on the green colour of Jade Mountain. One
described it rather fetchingly as if " … (it) looks like something done in
watercolour by an artist for very young children." It transpired that the
summit was neither jade nor grass but serpentine gravel that gave it the
distinctive verdant hue.

The discovery of Jade Mountain "solved the problem of the abundance
of nephrite implements scattered along the Bering Sea and the Arctic
shores. This isolated range, some 30 miles in extent, well above the Arctic
Circle, between the Kowak (Kobuk) and Noatak rivers, is 130 miles
inland from Kotzebue Sound."[4] Jade Mountain lies near the western end
of a belt of discontinuous ultramafic rocks, largely serpentinized for some
60 kilometres. Asbestos deposits associated with these serpentinites were
known from 1910, and during the Second World War renewed interest
resulted in some production of the tremolite variety of asbestos. The more
valuable chrysotile asbestos was also present, and both jade and asbestos
occurrences were examined and reported in 1945.[5]

In 1985 a more detailed examination indicated that the entire
region, known as the Cosmos Hills, was underlain by tabular bodies of
serpentinites as much as 130 metres thick and 8 kilometres long. They were
of Mississippian–Jurassic age, and overlaid the metamorphosed core of the
Hills. Jade occurred at the alteration zones between the serpentinite and
the schists and phyllites. Eons of erosion had released great numbers of jade
boulders, which ended up in the drainage creeks below.

Any outcrop of jade will shed blocks, of course, as erosion attacks the surface. The first pieces to be released are properly called talus blocks, and some may weigh many tons. Gravity will shift these large blocks as their underpinnings are worn away, but proper boulders are the result of stream action. Few streams can move a 20-ton block of jade, but smaller blocks may be transported by streams and glacial ice, and they may come to rest far removed from their source of origin.

In the 1960s commercial exploitation began with the formation of the Imperial Jade Company. From Jade Mountain in the west to Kogoluktuk River in the east, most of the creeks had jade float—sometimes up to 20 tons in size. Of course, not every boulder was of gem quality; typically, 5 percent met that criterion. And the extreme expense of doing business in an isolated location meant the costs of production made a commercial business difficult. Nevertheless, over the years several hundred tons of jade have been hauled out to Anchorage, principally by William Munz, and Ivan and Oro Stewart. Because the state does not keep records of production, reports published in magazines must be relied upon.[6]

It appears that Jade Mountain, followed by Dahl Creek 50 kilometres to the east, were the primary producers, and today Jade Mountain is owned and operated by Kobuk Valley Jade.[7] The company also sells Inuit carvings of walrus ivory and other handcrafted products. Its website advises that "you can watch while giant jade boulders are sawn and polished. Display cases contain hundreds of original and attractive Alaskan jade items such as rings, bolo ties, pins, earrings, necklaces, bracelets, cufflinks, and other items." Alas, there are no celts, picks, or axes. You can, however, purchase raw jade or just browse and enjoy free hot coffee. Lieutenant Stoney might have approved.

~ Washington: Continuing the B.C. jade story ~

Nature does not recognize the 49th parallel, so the extensive ultramafic geology that makes up the Fraser River gorge in southern B.C. extends into Washington. The Chilliwack Group, of Carboniferous and Permian ages, consists of greywacke, pelite, andesite, and basalt and is known as the Darrington Schist. Ultramafic bodies that are intruded into this formation are largely or partly serpentinized in contact with the schist, producing contact reaction zones that have led to the formation of jade lodes.

The earliest mention of jade in the literature appears to be from

Map 6: *The extensive ultramafic belts of British Columbia extend south into northern Washington, where a number of sites have produced jade.*

Harlan Smith,[8] an archaeologist with the American Museum of Natural History who joined the Geological Survey of Canada in 1911. He was well acquainted with jade artifacts and found many, plus raw material, in the Pacific Northwest. Since then jade float has been found on the beaches of Orcas, Vashon, and Whidbey islands—likely remnants of alluvial outwash from the Fraser River just to the north or the Nooksack River to the south. Inland, jade cobbles have been found in the Darrington area and around Deer Creek near Oso.

Close to the U.S.–Canada boundary, on Cypress Island in the San Juan Islands, an ultramafic body has been identified. The contact zone has been found, and although no nephrite has been identified, the presence of alluvial boulders on the nearby beaches suggests this is the source. At Twin Sisters Peaks, southwest of Mount Baker, an ultramafic body comprising largely unaltered dunite is serpentinized along the south side where the Nooksack River has cut into it, revealing numerous metasomatized tectonic inclusions within the serpentinite.[9] This site is located just south of the U.S.–Canada border. Although the visiting geologist did not report seeing nephrite, the numerous boulders and cobbles identified downstream in the Nooksack River must owe their existence, in part, to this source.

Washington has at least four *in situ* deposits, and there may be more. The best known and most productive was the Poor Boy Mine, owned in 1974 by the Washington Gem Jade & Mining Company. It was located on the southeast end of Mount Higgins (site 4 on map 6), 15 kilometres west-northwest of the town of Darrington, where serpentinite was in contact with the Darrington Schist and where both nephrite and rodingite formed. Blocks of as much as five tons were removed, but more commonly nodules of 10 to 20 kilograms were found. The deposit would require extensive bulldozer work to allow continuation of the operation. A number of creeks in the general area have yielded alluvial boulders, some reaching as much as two tons.[10] It would appear that the jade potential may be much greater than has been realized so far, with a number of ultramafic bodies worthy of careful prospecting.

South of Darrington, an *in situ* deposit (site 5) shows a 0.5-metre vein of serpentine on Helena Ridge, 10 kilometres south–southeast of the town, and there is likely more than one deposit in the area. A third lode (site 6) occurs on Cultus Mountain, some 12 kilometres east of the town of Mount Vernon. In 1953 James G. Stephens of Seattle came across nephrite and jadeite there while prospecting for steatite. It was unexpected to find both minerals, and the more so to find them *in situ*. The Stephens Mine would appear to be the only occurrence of the kind in the state. The identification of jade was confirmed by John W. Melrose of the Washington State Department of Mines. The deposit was worked from a short inclined shaft of about 10 metres—the lode was said to be over two metres thick and to lay between a hanging wall of slate and a footwall of serpentine, with jade occurring near the hanging wall. Other reports indicate that botryoidal nephrite has been found in the creeks in the area. The Sultan Basin (site 7), 50 kilometres east of Everett, shows contact along at least two reaction zones with typical rodingite minerals, so perhaps this is the site of the next jade discovery in Washington. Finally, nephrite is also reported to occur high on the west flank of Mount Stuart (which lies west of Leavenworth off Highway 2; site 8 on the map), on the Ingalls Peak ultramafic body.

~ Wyoming: The first finds ~

Until the major jade discoveries of British Columbia, the nephrite boulders found in the Lander area toward the Continental Divide in Wyoming were considered the most important. However, the colours have never been top-grade, instead tending to be brown–green, trending toward black. Black spotting is also a problem.

Early reports in the 1930s described cobbles and boulders found over a wide area, and at the close of the Second World War there was a minor jade rush similar to the one that occurred in B.C. in the late 1960s. Since the southern Wyoming region is treeless and subject to heavy weathering each winter, considerable material was found, with one report noting 436 pounds of apple-green jade collected in 1946. Curiously, however, there appears to be no evidence that the stone was worked by the Native people before European arrival. In the 1960s, five *in situ* deposits were identified,[11] and a boulder weighing close to 2,500 pounds from Wyoming is now in the Chicago Natural History Museum.

~ Oregon: More yet to be found ~

It may seem surprising that Oregon, lying as it does between the states of Washington and California, both of which have numerous jade occurrences, should have such a paucity of nephrite sites. Alpine serpentinites are present in the southwest of the state as an extension of the Klamath Mountains, and contact reaction zones have been

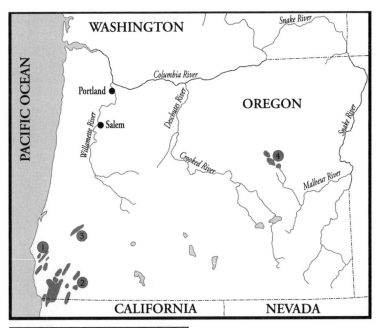

Map 7: *Oregon has a number of potential areas for nephrite formation, much of it concentrated in the south-west corner of the state in blueschist ultramafic bodies.*

reported.[12] Although nephrite is not mentioned specifically, the presence of rodingite does indicate favourable conditions, and nephrite is expected.

Similarly, the John Day locality on Baldy Mountain in eastern Oregon is reported to have metasomatic development along a serpentine–gabbro contact, suggesting that nephrite may be present. It is understood that the John Day rocks are the equivalent of the Cache Creek terrane in British Columbia, which may mean a positive mineral horizon. There is also a surprising lack of mention of jadeite in the state, despite there being blueschist rocks in north-central Oregon near Mitchell. There are also blueschists in the southwestern corner of the state, in the Klamath Mountains.

~ California: The curious cove ~

Although not prized by early immigrants to the Golden State (except for the Chinese miners, who do not appear to have found the stone), records show that both nephrite and jadeite were known as early as 1880. Jade pebbles of a fine green colour were found in a sea cave in southern Monterey County. Speculation at that time suggested that since jade was a form of currency in Mayan and Aztec cultures, the pebbles may have been hidden there from earlier times. The occurrence was only confirmed in 1936, when it was reported that "dark green pebbles and sand-eroded boulders of nephrite were picked up along the beaches of the Pacific Ocean from Plaskett Point to Cape Martin in southern Monterey County." However, it was only later that a bedrock source for the boulders was discovered at the base of a wave-worn sea cliff, thereby explaining the presence of the boulders found earlier.

Subsequent offshore explorations revealed numerous jade boulders of up to 3,000 pounds on the seabed, which were recovered during the early days of scuba diving. A report from 1958 describes an underwater cave the size of a room where the walls were pure jade.[13] Though not a particularly high-quality material, as was later determined, the combination of size and rarity must have made it an exciting discovery.[14]

There are numerous other jade outcrops documented in California, all of them small and with limited grades. Generally the nephrite deposits are found in the region between Monterey and Los Angeles, close to the coast, and indeed, several are shore-based. The jadeite deposits are clustered round the north-central region of the state, stretching from Sonoma County to Humbolt and Trinity.[15] An annual Jade Festival takes place each fall in Big Sur, Monterey County, celebrating the state's connection with the stone of heaven.

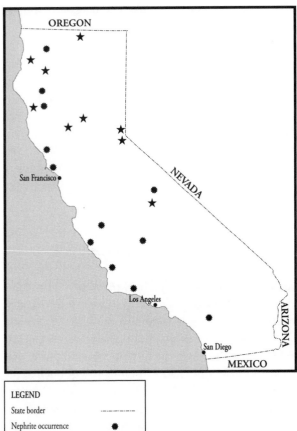

OREGON

San Francisco

NEVADA

Los Angeles

San Diego

ARIZONA

MEXICO

LEGEND
State border
Nephrite occurrence
Jadeite occurrence

Map 8: *The "Golden State" has numerous small deposits of both nephrite and jadeite, the best known being Jade Cove in Monterey County.*

~ Mexico: The pinnacle of jade culture ~

The jade of Mexico is (and was) predominantly, if not entirely, jadeite. Archaeologists have uncovered a wealth of artifacts from the remains of ancient civilizations that overlapped or succeeded each other, from as long ago as 1500 BC. Much has been written about these societies, and the significance of Olmec, Toltec, Maya, and Aztec cultures is well known. At the time of the Spanish conquest, the Aztecs were in ascendancy.

The amount of material needed to fabricate the vast number of artifacts recovered from all the known dig sites alone required a significant source of jade. The information on such a site seems to be missing.

In all of Mesoamerica there are just a few deposits known today. The question then arises, did the raw material come to Mexico via trade routes, and if so, from where? There are few accounts of *in situ* deposits within the country. The newly rediscovered deposits in Guatemala are filling in some of the blanks, and there is a possible site in Costa Rica, but the mineralogical differences in jades of various cultures and times suggest that there must be other sources.[16] The region is extensively overgrown, large areas are remote, and there is a strong possibility that old jadeite quarries may simply have been swallowed up by vegetation.

~ Europe: The West's cultural birthplace ~

During the Neolithic Age, both jadeite and nephrite were used in most European countries, primarily as axes, scrapers, and weapons. Although the early Europeans did not take jade culture to the high art that the Chinese, Olmec, and Maori cultures later did, there seems to be clear archaeological evidence that they were the first to do so. It's likely that the later development of a gemstone culture in Europe owes much to the earlier use of this stone.

Famous gemstones and gem jewellery were routinely recorded in early Greek and Roman literature, and birth gemstones and bridal jewellery evolved as a result. However, it was only in the twentieth century that the world's premier diamond seller, de Beers, took that gem culture to a new level with its successful "a diamond is forever" marketing strategy.

Jade tools were certainly used as early as 25000 BP.[17] In 1921 a severe drought lowered the level of Lake Geneva in Switzerland to reveal a community built on stilts out over the lake, and a plethora of jade tools were found amid the submerged artifacts. Elsewhere, on the Lake of Bienne, a stilt village extended over three hectares (six acres), complete with drawbridge and causeway connecting it to shore. At that locale, over a thousand jade implements were recovered.

It is clear that in Neolithic Europe both nephrite and jadeite were the tool-making stones of choice. In Italy people were using jade axes as late as 2500 BP, a millennium after the start of the Bronze Age.

Nevertheless, with the arrival of bronze, jade use naturally declined and all but disappeared until the return of Christopher Columbus and other explorers from the New World and the Portuguese from Cathay (China) in the sixteenth century.

During the nineteenth century there was considerable debate as to the origin of the jade artifacts found in Europe—both jadeite and nephrite tools had been excavated from early sites. A noted expert, Dr. L. Heinrich Fischer of Freiburg im Breisgau, Baden, Germany, collected and wrote extensively on the subject. He argued that jade was not, in fact, indigenous to the continent but had been "dropped or lost" by prehistoric tribes.[18] This, in turn, suggested an early trade route between Europe and Asia. Fischer's theory dominated the latter part of the century, despite opposition by Dr. A.B. Meyer of Dresden and others, who argued that sources of jade did exist in Europe but had either been depleted or were lost to memory and would in time be rediscovered. It was therefore with a certain sense of irony that *in situ* nephrite was found in 1885 and documented by the noted American gemologist George Kunz at Jordansmuhl, Poland, within 300 kilometres of Dr. Fischer's home.[19]

Other finds have subsequently been revealed. In Italy nephrite deposits were discovered in 1906 near Monterosso and, later, at Sestri Levant on the coast. In Switzerland, *in situ* material occurs near Salux close to the Lichtenstein border and farther south at Poschiavo, close to the Italian border. As for jadeite, a source of that was finally found too, at Piedmont in the Italian Alps in 1903[20] and later in Serbia, Yugoslavia.[21] And not too far away, a well-known jadeite deposit has been worked in the Russian Polar Ural Mountains for some time.[22]

Hunting the Stone of Heaven: A Personal Odyssey

1986: Labrador — The Smithsonian adventure

In 1829 an obscure Englishman named James Smithson died in Italy, leaving behind a will with a peculiar footnote. A life-long bachelor, Smithson left his entire estate to his nephew, with a caveat that if the nephew died without an heir, the money would go to the newly established United States of America, to found, in Washington, "an Establishment for the increase and diffusion of knowledge, that Establishment to be called the Smithsonian Institution."

Smithson had been a Fellow of the Royal Society of London and had published some work in geology. He would later, after his death, have the mineral "smithsonite" named in his honour. Six years after his death, his nephew, Henry James Hungerford, died without children, and in 1836 the U.S. Congress authorized acceptance of the gift from a man who had never set foot in the Americas.

President Andrew Jackson sent diplomat Richard Rush to England, and two years later Rush set sail for home with 11 boxes containing a total

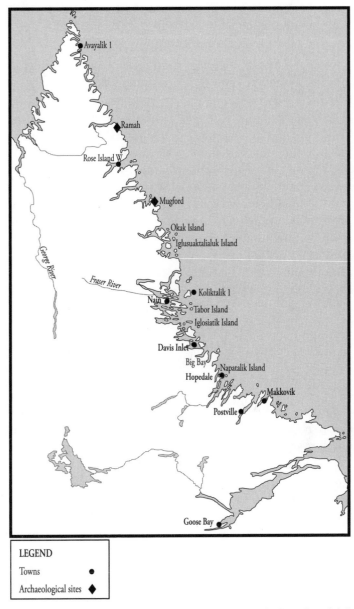

Map 9: *The early occupants of Labrador used nephrite tools, but where did they come from? In 1986 the Smithsonian Institution invited the author to join an expedition to try and trace the source.*

of 104,960 gold sovereigns, eight shillings, and seven pence, as well as Smithson's mineral collection, library, and scientific notes. After the gold was melted down, it amounted to over half a million dollars—a small fortune in those times. In 1846 the establishment of the Smithsonian Institution was signed into law by President James Polk.

Today the Smithsonian comprises 18 museums, plus galleries and research facilities in the U.S., as well as internationally. There is the National Museum of Natural History, the National Zoological Park, the National Museum of American History, and the National Portrait Gallery, to name just a few. The National Air and Space Museum is another and has the distinction of being the most visited museum in the world.

At the Museum of Natural History, Drs. William Fitzhugh, Christopher Nagle, Steven Cox, and Susan Kaplan studied the early occupants of the Arctic and found that the times in which the inhabitants lived could be divided into three consecutive periods. These were Early Dorset (2500–2000 years BP), Middle Dorset (2000–1400 BP), and Late Dorset (1000–800 BP). Because these were Stone Age people, they left almost no archaeological records apart from the stone they worked and used. One method, therefore, of tracing migrations was to excavate early sites, analyze the stone, and try to determine where the source of that stone was. In this way, anthropologists might deduce where the predecessors of today's Inuit travelled and traded.

Much of the stone they used was chert, which is similar in properties to the flint found in Europe. Chert can be readily chipped into complex shapes, holds a sharp edge (important for knives and arrowheads), and is reasonably available—notably at Ramah on the Labrador coast, close to 59⁰ N, and to a lesser extent at Cape Mugford south of 58⁰ N on the same coast.

A paper by the Smithsonian's Dr. Chris Nagle set out to trace these migrations, based on the evidence of chert found at various sites. It is interesting to note that fragments had been discovered up to 400 kilometres away from their source, showing fairly dynamic trading patterns by the peoples who used them. However—and this is where I became involved—the evidence also showed that there were nephrite artifacts too. Where had they come from? There was no known jade source on the east coast apart from one small site on the northern tip of Newfoundland at Noddy Bay, but the material did not match. Further, a single jade cobble had been found at Saglak Bay in northern Labrador, but that source was also unknown.

The current school of archaeological thinking suggests that the Arctic was populated in a series of human waves from the west (Alaska), as the last Ice Age receded. There was certainly jade in Alaska. Had the early Dorset people brought it with them? Or had they merely brought the *knowledge* of jade with them and instead found a source along the way? If so, where was it?

Sir John Richardson, in his account *Arctic Searching Expedition,* published in 1851, mentioned that jade had been found in what came to be known as the Rae River. The site was described in some detail but has not been found again, despite repeated expeditions to the area. Perhaps it had not been jade at all; in 1851 there was still a lot of confusion about the properties of the stone.

It was interesting for me to learn that the Inuit were not the first people on the Labrador coast. The ancestors of the present Montagnais–Nasapi Indians had been there for as long ago as 7,000 years, but as far as we know they did not use jade. However, it was the belief that a local source did exist somewhere along the coast that led Dr. Nagle to undertake a geological investigation, in tandem with other programs being conducted by Dr. Fitzhugh.

The first challenge was obviously to collect as much local geological information as possible, and so Dr. Nagle approached the Geological Survey of Canada. They put him in touch with Dr. Ingo Ermanovics, who had spent several field seasons in the general area and had produced geological maps and reports. With additional verbal communication, a number of potential ultramafic locations were identified.

It was against this background that I received a letter from Dr. Nagle in 1984, offering me a chance to look for jade with the Smithsonian's ongoing exploration in Labrador. As I was retired and no longer actively engaged in the geological investigation of jade occurrences, I was pleased and honoured to accept the invitation. And since the institution offered to pay all expenses for the month-long expedition, I was also quite satisfied with the arrangement. In due course I left the Okanagan Valley, just as the cherries were ripening, and some considerable time later arrived in Goose Bay, Labrador. It was July 11, 1984.

There were a few days' delay in Goose Bay, during which time the group arrived and introduced themselves. Heading the field operation was Dr. William (Bill) Fitzhugh, a senior researcher at the Smithsonian and later the Director of the Arctic Studies Centre at the National

Museum of Natural History in Washington. He had been going to the Arctic every summer for nearly 20 years. The other members of the group were Bill's son Ben, the archaeologists Susan Kaplan, Steven Cox, and Chris Nagle, plus Robin Goodfellow, who was a Canadian taking a leave of absence from the Canadian Broadcasting Corporation to come along as a field assistant. She was especially welcome, as she had a share of a house in Goose Bay, which we used as the base of operations.

Our ship, the old RCMP patrol boat *Tunuyak* (meaning "caribou back fat"—well, that is what I was told). was designed for two constables and one prisoner. Having seven on board was going to be a challenge. As the boat wasn't ready, there was some time in Goose Bay–Happy Valley.

The double place name derives from the history of the town. During the Second World War, the U.S. Air Force built Goose Bay as a staging ground for air transport across the Atlantic to Britain. Later the facility was handed over to the Canadian Forces but still remained a military base. As a result, the town of Happy Valley was developed for the civilians who needed a place to live off-base.

Why it was called "valley" is a mystery, as it is built on a vast alluvial sand plain formed by the outwash of melting glaciers during the Pleistocene Ice Age. To the keen-eyed observer, a ring of minor hills encircles the area. These are referred to locally as "mountains"; hence, perhaps, the appellation "valley." It was the building sand, plus its distance from coastal sea fog, which made the place attractive as an airfield, as the coast of Labrador is renowned for its fog and bad weather, even in summer. As well, the flying pests have something of a reputation—local anecdotes advise there are two kinds of mosquito: "The small ones fly straight through the mesh on the windows, and the large ones kick in the door. And the no-see-ums get back up after you hit them with a baseball bat."

Labrador became a British possession in 1763 under the Treaty of Paris between Britain and France. Later, in 1809, it merged with Newfoundland, finally joining Canada in confederation in 1949. Mineral collectors will know that Labrador is famous for the gemstone *labradorite*, a plagioclase feldspar that exhibits brilliant flashes of blue, yellow, and occasionally red "fire" somewhat similar to opal. The locality at Tabor Island near Nain is a well-known labradorite source and was operated as part of an Inuit co-operative. However, when we visited, we found it inactive, although there was an abundance of material on hand. Apart from Tabor, there were many other localities in the vicinity, but perhaps the demand for the mineral, plus the costs associated with extracting it, made it prohibitively expensive when compared to the sources in Finland and Madagascar.

With the *Tunuyak's* fuel pump still inoperative as of July 15, five of us flew to Hopedale, leaving Bill Fitzhugh to bring the ship up later. Transport was a single-engine de Havilland Otter on floats, owned by Labrador Airways. A small plane cruising at relatively slow speed a few thousand feet up is a fine way to travel if time is not pressing. As the plane rose from the surface of the lake, we could see the extent of the sand plain, ringed by hills, which to the southeast are called the Mealy Mountains (450 metres high). The terrain below, pitted with lakes and spruce swamps, plainly showed a southeast trend imparted by the glaciation of more than 10,000 years ago. Landforms were clearly the work of ice and the meltwater of the waning phase. Long, snaking eskers (ancient lateral moraines) were numerous, with water everywhere. The farther we went north, the more bare rock was exposed, and on approaching the fiords running southwest from the coast, patches of snow appeared. In July? It was a barren but beautiful land.

On the final approach to Hopedale we saw lots of sea ice and icebergs. The plane taxied over to the dock, where local Inuit fishermen were mending boats or were busy in the fish-processing plant. Straight ahead, up the boardwalk over the bare and ancient rocks, were the buildings of the Moravian Mission, our accommodation.

The Moravians have been in Hopedale since 1777, and at one time theirs was a thriving community with a trading post and big buildings for residences, school, and infirmary. At present, many of the mission buildings are unused. We all spent time viewing the rarely visited museum at the mission, which had a collection of early memorabilia, and it being a Sunday, we were invited by the Reverend Case to a service, where we constituted about half of the congregation. Afterward John Case laid on a lunch of delicious smoked char.

We spent the following few days waiting for the rest of the party to arrive by sea. There was no heat or water in the big building where we stayed, so although we lacked nothing in space, we were obliged to go over to the Case's house for the time we were there. While waiting for the arrival of a 20-foot freight canoe and an 18-HP outboard to come down from Nain by coastal packet, we had time to look around—it didn't take long to see the town. There were about 500 inhabitants, mostly Inuit, and the scrubby trees were outnumbered by the power poles. Above the town, remains of a radar station marked the eastern extremity of the American DEW Line of the '50s and '60s. The radar site was a mess, with no effort made at any form of reclamation.

Sometime during our second night the equipment and canoe were dropped off. We had been warned that the canoe needed some work, and it certainly did. Chris went up to the store for fibreglass, paint, and nails, while I sat on the dock, trying to understand the pull of a place like this but failing: one must probably have to have been born here not to see the disadvantages, discomforts, and isolation. Small fishing boats with outboards on the back were everywhere, with some of the larger vessels sporting cabins for shelter. On the dock children fished constantly, and there seemed to be no shortage of sculpin and codfish caught on their jigs.

The following morning we launched the repaired canoe, and although it leaked slightly, it was certainly seaworthy. After lunch we gassed up the motor and took off south of town to a serpentine locality near the top of Adlatok Bay. These are the kind of host rocks that, in B.C., might mean jade. However, we found nothing, although it was not an exhaustive search, and we returned to Hopedale for one more comfortable night. It rained heavily, so the next morning we bailed the canoe and then headed south to Adlatok Bay again, stopping en route to look at serpentine deposits but finding no jade. We continued to the bottom of the bay, found a suitable campsite on an island at the mouth of a salmon river and tried a few casts, but the fish weren't biting.

Soapstone had been reported on this particular island, and eventually we found a small quarry, more or less obscured by low birch and alder. Back at camp it started to rain and continued heavily all night. Welcome to summer in Labrador. The following morning we tackled the quarry again, the archaeologists measuring and excavating, and I hoping to find some evidence of jade, but no luck. It was evident that a considerable amount of soapstone had been removed over the decades—possibly as much as 500 tons. Narrow cuts in the blocks suggested modern quarrying techniques, but it was obvious there had been no activity in recent years.

We returned to Hopedale to hear that our ship was finally on its way and would arrive on July 20, so in the meantime we headed north for Napatalik Island. All the way there, the two white radar domes beyond Hopedale stood like sentinels behind us, giving a curious and misplaced sense of security in an otherwise almost featureless land. Without them,

I'd have been lost in the maze of channels, reefs, and islands. During the trip I sat in the middle of the canoe, like Governor Simpson with the *voyageurs* speedily paddling me on my way. Looking back in my notes, I had commented that in Labrador, time seems to stand still (apart from the radar domes on the horizon).

We passed a few small ice floes, saw a seal momentarily surface and through the cold air witnessed a large storm brewing offshore. Once at Napatalik Island we cruised along its western shore toward a smaller island (whose name I neglected to note in my diary) and landed in a sheltered anchorage, keen to stretch our cramped legs. On the horizon the radar domes were still visible, but we felt we were in the middle of nowhere as we set up our tents—it was after 9:00 p.m., overcast, and threatening rain. I recall that the mosquitoes were voracious.

The next morning we decided that because the sea was so calm, we should visit some locations on the mainland to the west, despite being aware of soapstone on the island we were on. This we duly did but found nothing of significance. However, it was a great day, the kind one remembers as making everything worthwhile. The colours around us were brilliant, the light sharp, and the air so clear it seemed we could see over the horizon. But what a dearth of humanity! Occasionally one might see or hear a passing fishboat or see vapour trails of jets bound for the great cities of Europe, but on that empty coast was a solitude that seemed to encompass all creatures. We saw few birds and fewer mammals, and the desolation seemed total, even though below the sea's surface there was no doubt a plethora of life.

On the following morning some of us walked to the north end of our island for a closer look at the soapstone deposits. At the shore was plenty of evidence of modern Inuit—plastic was everywhere: bottles, tubes, sheets, old boots, etc.—but stone rings marked the tent sites of more ancient camps. On the beach were two large boulders of soapstone, with others strung out in a line along the narrow zone of ultramafic rocks that gave rise to them. One boulder had the beginnings of a rectangular block chipped out of it, destined to be removed and hollowed out at some convenient camp in the future.

We headed to a new site on the mainland again the next morning. Once more, though, it turned into an unsuccessful jade hunt; there were minor amounts of talc, but no serpentine. We left Steve Cox at a previously known archaeological site where, we had been told, one of the Morovian missionaries had been murdered. The site turned out to be Mid-Dorset (about 200 AD). While Steve explored, Chris and I went off looking for rocks more favourable to jade formation but found little of interest.

We were heading back to camp when we learned via radio that the *Tunuyak* was on its way from Hopedale, and when it arrived at about 8:30 p.m. that evening, we all went on board to celebrate. There was a new addition to the crew: Larry Jackman, a freelance journalist, was gathering firsthand information on the Smithsonian's activities. He was only on board until we land-campers rejoined the party, at which stage he was heading back to the world of hot showers and central heating.

The next day we headed for Windy Tickle—on the west coast of Canada we'd say Windy Channel—to drop off Larry for his flight back to Goose Bay. While at Windy Tickle we received an invitation to dinner from the captain of the *Maria Teresa*, a Portuguese fish-buying boat. Our standards were rising again! But as we pulled alongside the *Maria Teresa*, all we saw was an extremely rusty hull and decks heaped high with mountains of salt. We wound our way up to the bridge, and as we entered, any idea that this was a tramp steamer disappeared. The bridge was crammed with the latest electronic equipment for communication and navigation. Both the captain and first mate spoke English quite well, so there was no problem with communication, and in due course a bottle of vermouth was produced and we adjourned to the warmth of the wardroom. There we enjoyed a well-served, fine meal, and not surprisingly, the entrée was salt cod. There followed many toasts in wine and brandy, to as many famous Portuguese as we could remember.

We left Windy Tickle reluctantly, just as bad weather moved in, and were obliged to hole up in the lee of a number of islands to shelter from the blow. I had left my sleeping bag back at our island by Napatalik, so while on the voyage I had used the only spare blanket aboard the *Tunuyak*. I was not comfortable, but it was no doubt better than camping out. In poor weather, the *Tunuyak* was certainly an improvement over the canoe—we could stand up in the wheelhouse and astern. However, it was not a fast vessel, so voyages of even a few miles could be lengthy. On calm days, it was very pleasant to sit on the engine hatch and read, snooze, or admire the passing scenery.

From the onset of the storm, though, it was a full three days before we made it back to our camp on the island. On our arrival we found the tents

The 1984 Labrador team. Back row: jade geologist Stan Leaming, archaeologist Dr. Chris Nagle, CBC reporter Robin Goodfellow. Front row: expedition leader Dr. William Fitzhugh, archaeologist Dr. Susan Kaplan, archaeologist Dr. Steve Cox.

down and the sleeping bags soaked. However, there was no permanent damage, and the group soon began their archaeological work.

There were numerous sites in the area, most of the Mid-Dorset era, and as there was little chance of finding jade in the bedrock, I went prospecting for other interesting minerals. Subsequently a group of us left Steve, Robin, and Chris working and went back to Hopedale for supplies, after which we set out for the Okak area farther north, passing through Windy Tickle again. The weather had stabilized, and I spent a perfect time atop the wheelhouse, with superb views in all directions. We passed the lone house of a summer fishing camp, built on a low terrace close to the shore. Behind us, beautifully banded and faulted, was gneiss of the Archaen era—some of the oldest known rock in the world, dated to the unimaginable age of 3,000 million years.

We motored out of the tickle and made a "hard-a-port," as the sailors say, on the way to Big Bay, where we stopped to check a windblown patch of sand, called a "blowout" by archaeologists. These are great places for the discovery of artifacts, provided there have been inhabitants above the site at some stage. This particular blowout was previously known to yield artifacts made by Indians of the Maritime–Archaic culture. Not too much was found during our short stop before we turned west into the setting sun.

On the following morning we were away early, running for the head of Big Bay. On the charts it looked small, but from the deck of a ship it appeared enormous. It was indeed a big bay. There was a possibility of jade-bearing rocks in the region, so we anchored and went ashore. The geology was deceptive, however, and we found neither jade nor even serpentine. Disappointed, we embarked once again, and set a course for North Tikagakyuk Island, stopping briefly en route at other Maritime–Archaic sites. We ran then for Davis Inlet, where we soon saw the original village and the one-time Hudson's Bay post, which, I was informed, was the first of its kind in Labrador.

The present house there is only 60 years old but already showing the ravages of weather and neglect. At the time of our visit, it was the summer home of the Saunders family. Bill Fitzhugh was well-known to them, and the Saunders were clearly pleased to see him again. In the sheltered inlet surrounding the house was quite a stand of timber. First seen by the great Arctic explorer John Davis in 1585, it has probably changed little. The town of Davis Inlet was about five kilometres west of the house and was the centre of the Mushuau Innu, with about 250 people living there at the time. In 2002, the entire town, now home to about 650 residents, moved 15 kilometres to the newly constructed Natuashish. We called in briefly to check for mail before heading off in search of a sheltered anchorage for the night.

The next morning (July 29) we landed at our destination, or so we supposed: the island of Iglosiatik, where numerous sod house sites were reported. It transpired we had landed at the wrong place, and the site we were really looking for was about eight kilometres to the west. I decided to walk, while the rest of the crew returned to the *Tunuyak* and motored round to the correct location. My hike was pleasant but of no value archaeologically or geologically speaking, and when I arrived the others were busy digging test pits in some of the depressions that marked the area.

My colleagues informed me that the site had been inhabited just 400 years ago (about the time of John Davis) and had been one of the largest settlements along the coast. Susan Kaplan worked with a metal detector to verify any contact with Europeans. The archaeologists were all quickly engrossed in their work, where the next trowel of dirt might reveal some long-buried artifact, shedding light on the origins of the mysterious people who chose this corner of the continent.

Over the years, Stan visited the jade properties in Lillooet, B.C., many times, using everything from a four-wheel drive to a motorbike. One of his best excursions was on one of Patricia McEwen's horses; they reached the upper part of the Shulaps Range.

This is a typical jade cobble that has been tumbled to a shiny finish by river action. It was over a century before discoveries like these in the Bridge, Yalakom, and Fraser rivers in B.C. led prospectors to their sources in the Shulaps Range in the mid-1960s.

This 30-metre-high waterfall on O'Ne-ell Creek in B.C. blocked access to a higher area where the "Jade Mountain" was later discovered.

Stan pauses for a picture on the south side of King Mountain near Provencher Lake, B.C., during the mapping of ultramafic belts in the Dease Lake–Cassiar area in the early 1970s. The summer days are long here, but the season itself is short.

King Mountain near Letain Lake, B.C., seen from the north. The once-glaciated valley in the foreground proved to be the world's most prolific source of nephrite boulders, which were scattered widely across the tundra after the ice receded 10,000 years ago.

Walter Ellert's operation near Provencher Lake, B.C., is shown here. At the 2,000-metre level, a vein of nephrite zigzagged across the slope for about eight kilometres, and was known as the "China Wall."

Mohawk Oil's Cry Lake Jade Ltd. was the first well-funded corporation to get into the Dease Lake, B.C., nephrite business. With large earth-moving equipment, bigger jade boulders could be processed using a combination of hydraulic tools and crawlers.

Drilling a boulder on the Northern Jadex property in B.C. Nephrite is made up of interlocking rock fibres, making it one of the toughest natural materials in the world, so cutting and drilling take time. Diamond-impregnated blades and bits have helped, but the process is still slow.

Rodingite, sometimes known as "white rock," shows a contact with serpentinite at McClure Creek, near Dease Lake, B.C. Like serpentine, rodingite is an indicator mineral, its presence allowing the small nephrite lode to be found in an otherwise complex geological matrix.

The Tunuyak *is loaded in preparation for leaving Nain, Labrador. In praise of the vessel, Stan noted that in poor weather, the* Tunuyak *was better than a canoe.*

On the treeless coast of Labrador, the team of specialists camped close to known archaeological sites, where excavations were carried out. Earlier research had shown that some cultures had used jade implements, so Stan's reason for visiting the sites was an attempt to find the source(s). The expedition vessel Tunuyak *can be seen to the right of the yellow/red tent.*

In China, not far from the Karakash River, the road cut through a strikingly formed rock arch of Carboniferous conglomerate. In a wet climate, a feature like this would have a shorter existence. Here, in the rain shadow of the Kunlun and Pamir mountains and in the acutely arid desert environment, it could survive for centuries: only the wind would erode it.

On the hunt for jade in Siberia, the team transferred to a pair of tracked vehicles, which they met on the banks of the Kitoi River. Although it was mid-summer, the valleys were routinely filled with clouds in the mornings. The dampness did not help the reliability of the vehicles, which had to be repaired almost hourly.

Alf Poole relaxes against the bumper of the ex-Russian Army truck that took the team to the jade fields southwest of Irkutsk in Siberia. A surprising amount of beer and vodka was packed into its cavernous back.

Kirk Makepeace, CEO of the world's largest jade-mining operation (based in B.C.), wets the sawn face of "Polar Pride," a nephrite boulder from Serpentine Lake of such outstanding quality that it has been described as a "once-in-a-millennium discovery."

Harbour Play *by Lyle Sopel, 18" x 11" x 10", jade and black quartz, shows a sea lion's sinuous form against delicately swaying sea grass.*

Kalamalka Reflections *by Deborah Wilson, 72" x 53", stands in front of the Vernon, B.C., library and pays homage to nearby Kalamalka Lake, the "lake of many colours." Water plays on the sculpture from the base, a 500-kilogram jade boulder in the middle of an L-shaped pool. The work is lit with fibre optics.*

Sitting off to one side, watching their efforts, I turned to the sea, where our landing boats were moored. Because of the rocky shore and the changes in the tide, it was customary to anchor them off the rocks and bring a line ashore. It occurred to me that, in this case, these were also *life*lines: if the rope should break and the boats drift offshore, we would all perish from exposure or hunger. The marine traffic in the area was non-existent, and the chance of a passing fishboat spotting a band of stranded explorers was slim.

During my wanderings in the vicinity of the archaeological diggings, I occasionally came across a combination of circumstances that permitted me to have a bath. One requires, of course, the melting snow to be warmed and a natural bathtub to be strategically located. Since freezing water is not conducive to leisurely soaking, I looked for pools that were not too deep so that what little warmth there was in the sun would elevate the water to some reasonable degree. The next consideration was the relative absence of mosquitoes. I also required complete immersion, and finally, my natural modesty demanded some form of privacy.

This combination was not easy to find, and now and again I was forced to tolerate less-than-ideal conditions. It would be no surprise if the feeling of cleanliness had little appeal to people living in that environment, and I more than once wondered what it would be like to go for months, if not forever, without being free from smell and dirt. What must it have been like to crawl into a sod house, where a family lived with their food scattered around, fresh hides drying on the walls and smells from who knows what else? I came back to earth with a jolt. Susan had found some iron, and there were bones from marine mammals in the dig.

Later we ran down the east side of Iglosiatik for a night's shelter, and the following morning we went ashore in search of another Maritime–Archaic occupation. On the upper beach bits of Romah chert were everywhere, even a limited amount of Mugford chert. There were also millions of mosquitoes. It was one of those crisp days when images and photographs are sharp and clear. Numerous stone mounds marked the surrounding gravesites, although the archaeologists did not disturb them—not for lack of interest but because of sensitivity to local opinion. Apart from this, they took great pleasure in excavations and often spent days working with care on a single small area. Unfortunately, we did not have enough time to stay at the site as long as we would have liked.

Late that same afternoon we left for Kamarsuk, where Steve Loring

Stan walking up to the labradorite quarry on Tabor Island. It was from this site that the famous gemstone with the blue and yellow fire was mined for export. Under the jurisdiction of the local Innu since 1984, the site showed no evidence of being operated.

was doing some independent archaeological research. There was a good anchorage there, so we spent the night, but early the next morning ran north to Nain.

The sea was calm, and we cruised through an archipelago of islets. There was more standing timber here, and the land looked less bleak than it had farther south around Hopedale and with less snow too. Could this be the start of summer? On the way we stopped at Tabor Island, which lies about 30 kilometres south-southwest of Nain. As mentioned earlier, this was one of the original labradorite (feldspar) gem sites, but it was clearly not being operated.

On August 1 we motored into a cove in Snyder Bay, not far from a fishing camp. To the south, the Kiglapait Mountains formed an impressive backdrop as Bill Fitzhugh cooked up a mass of scallops. There was a cool northerly wind, and it had clouded over again, but our anchorage was secure.

I went ashore to have a look at the Snyder Group of rocks, which were atypical ultramafics with quartz and marble beds, but there seemed to be no great potential for jade there. That, at least, was my opinion in 1984. Now I am not so sure. A local Innu appeared from a nearby fishing camp, wanting to know if this was the same group who always looked at the rocks! That night (I note from my diary) there were no mosquitoes, and I slept warm for the first time in awhile—at 67 years old the circulation isn't always the greatest.

The following morning dawned overcast, but there was no wind in the shelter among the small islands when we arose at 7:00 a.m. From sea level the Kiglapait Mountains looked like real mountains, although they are probably barely 1,000 metres in height. On Iglusuaktalialuk Island (*iglu* means "big island with sod houses"), I went ashore to look at some ultramafic rocks that had been reported on the east side. It turned out they were small bodies, with some talc and biotite along one contact, but nothing encouraging. Walking back past an Innu fishing camp, we noted three tents, and sharp-eyed Chris Nagle picked up a nephrite celt on the trail where local Inuit must have walked many times. Otherwise the land seemed empty, apart from a few ptarmigan, a whale or two in the bay, and a small dog who tried to follow us from the fishing camp.

We later left, heading north again, this time for Moore's Island and a reported soapstone outcropping. In among the Okak Islands the wind started to pick up, so we ducked into Moore's Harbour, where we spent the entire day of August 6 waiting for the wind to ease. When it finally did, we covered Norton Island and Orton Island (where I found lots of anorthosite and labradorite but no serpentine), and the ominously named Skull Island, where we landed for archaeological purposes.

The weather, never good, deteriorated again and we turned for Nain. It was the end of the season and time to head home. Curiously, while on the boat and busy with projects, there was no real sense of discomfort, but the moment the bow swung south in a homeward direction, the cabin seemed crowded, the food bland, and the vessel's speed snail-like. We couldn't get back fast enough.

Four days later most of us were in Goose Bay–Happy Valley, saying our goodbyes and taking different planes to different locales across the continent. My jade search was over. I had known from previous research that soapstone localities were fairly numerous along the coast and islands. Some sites had been quarried by ancient peoples, but at none of them, it turned out, did we find any indication of jade. In fact, few showed any sign even of serpentine.

In the few localities marked *calcsilicate* on Ermanovics' maps, we were also unsuccessful. The only jade found was in a few of the archaeological sites excavated by other members of the team. Naturally I was disappointed, in that I had come 5,000 kilometres to participate in an exploration. It was hard on my ego, as the so-called "jade expert." In retrospect, I wonder if I might have missed deposits in the few places that seemed propitious or if we simply missed the places where there was jade. With the benefit of hindsight there may be an ultramafic suite in the rocks in the vicinity of Florence Lake, some 80 kilometres south-southwest of Hopedale. We did not visit the area, but Ermanovics reported *tremolite* in serpentinite layers several metres thick, exposed for some eight kilometres and up to 300 metres wide. In size at least, the belt is comparable to those in B.C.'s Cordillera.

1987: China — Where jade began

In a word association test, "jade" is likely to evoke the response "Chinese," or, less often, "green." Given the four-millennium history that China has had with this mineral, "Chinese" is appropriate enough, but "green" is not. Not all jade is green. Nor is jade confined to China or to the Chinese, even though the material has been in their tradition for a very long time. But let's begin with gold.

Gold has long been a much sought-after metal, and when erosion liberates it from the bedrock source and deposits it in suitable water courses, it becomes pretty easy to recover. Even an old frying pan can be used, although properly designed gold pans are a great deal better. That, plus a shovel, constitutes the machinery for alluvial gold mining. All you need to do is shovel a little bottom material into the pan, add water, and swirl the mixture around to wash out the fine sand and mud, then pick out the pebbles, and there is the gold.

Back in 1848 one James Marshall and his crew were widening the tailrace for the new Sutter's Mill on the American Fork River in California, when Marshall noticed a few glimmers of metal in the mud. Thus began the gold rush of the forty-niners, for by the year 1849 the whole world knew of the discovery. The news reached China soon enough. Many Californian mines, especially in the south, were worked by immigrants, who came solely for the gold. Chinese, Chileans, Mexicans, Irish, Germans, French, and Turks all sought their fortune in the sunshine state. However, unlike their American-born counterparts, most foreign miners had no intention of staying in California. Their goal was to find the gold and then go home.

A decade later gold was discovered again, this time in the Fraser River. In 1858 the first Chinese gold miners came from San Francisco to what would later become British Columbia. There they joined thousands of other prospectors in the trek northward along the Fraser Canyon. Still others came to Canada directly from Guangdong Province in southern China.

Their arrival marks the historical establishment of a continuous Chinese community in Canada. To this day Canada is known on the Chinese coast as "Gold Mountain." The historical significance of the Chinese placer miners was their association with the Asian tradition of jade lore. When the green rock showed up in their sluice boxes on the Fraser River, they saved it to be shipped back to China. By a nice twist of fate, a century later it was the Asian jade buyers who first came to British Columbia to select raw material for the jade-carving industry in Taiwan and mainland China.

The resurgence of interest in jade in the 1960s can be attributed initially to the entrepreneurship of Taiwanese businessmen. Their own mountains contained jade but, with increasing costs, it had become cheaper to import Canadian jade than extract the product from the underground mines near Fengtien, on the east coast of the island. Since 1980 there has been little production in Taiwan. Before 1985 total production amounted to 11,000 tons, most of which went into small carvings and jewellery items, mainly for the tourist trade (both local and foreign).

Across the straits, the People's Republic of China has some very large carving factories, which depend on imported material of many kinds. The limited amount of jade gathered from the alluvial deposits in Xinjiang Province ended up in Chinese factories where, along with lapis, rhodonite, tiger-eye, rose quartz, jadeite, and serpentine, it fuelled a growing local and international market for small carvings and low-cost jewellery.

In 1986 I was privileged to visit the People's Republic of China, and during my stay I was able to see the famous jade streams in Xinjiang Province. The local population is made up primarily of Uighurs, a Turkic-speaking Muslim community, where many are farmers who earn a sparse livelihood in the irrigated fields along the famous jade rivers, the Urungkash and Karakash. In the off-season they collect jade cobbles along the banks and gravel bars, and sell them, for the most part, to Chinese buyers from the eastern jade-carving factories. It's worth noting that Chinese in ancient times preferred white jade to the more typical and abundant green jade. We know that white jade does occur in Siberia, northeast of the north end of Lake Baikal, and it has been speculated that this material found its way to the workshops of the Chinese artisans in Beijing, Shanghai, and elsewhere.

Prelude to an expedition

It may seem a long way from New Zealand to China, but an interesting connection had developed that led to my trip. In years past, the Maoris' enthusiasm for jade had rubbed off on their European invaders, and among the descendents of the early British farmers was Russell Beck, a jade collector who was also director of the Southland Museum and Art Gallery in Invercargill. One of his associates learned of my interest in jade and invited me to come along with them on a trip to China in 1986.

The jade mines in the Kunlun Mountains of western China are the most famous in the world, for they are the source that has fed that country's passion for the stone. The rivers on both sides of Khotan (also known variously as Kotan, Hotan, Quotan, Cotan, Yu-thian, Ilchi, Yotkan, Hotien, and Hetein) are named after jade, and it is there that boulders, cobbles, and pebbles of nephrite have been found for at least 2,500 years.

The traveller Chang Juang-yi noted in his diary (936–943 AD):

"Every year in the fifth and sixth months a swollen torrent of water rushes down, and the jade follows with the current, its quantity depending on the size of the flood. The water recedes during the seventh and eighth months, and it can then be collected, the jade being fished for, as the natives say."

Once found, the jade was bartered and collected before its long journey east to the fertile plains of China. Sir Henry Yule noted:

There is no article of traffic more valuable, or more generally adopted as an investment for this journey, than lumps of a certain transparent kind of marble which we, from poverty of language, usually call jasper. They carry these to the Emperor of Cathay, attracted by the high prices which he deems it obligatory on his dignity to give; and such pieces as the Emperor does not fancy they are free to dispose of to private individuals. The profit of these transactions is so great that it is thought to compensate for all the fatigue and expense of the journey. Out of this marble they fashion a variety of articles, such as vases and brooches for mantles and girdles.

There are two kinds of it; the first and more valuable is got out of the river of Cotan, not far from the capital, almost in the same way as divers fish for gems, and these are usually extracted in pieces about as big as large flints. The other and inferior kind is excavated from the mountains. That mountain is 20 days' journey from this capital (i.e. Yarkand). The extraction of these blocks is a work involving immense labour, owing to the hardness of the substance as well as the remote and lonely position of the place.[1]

Xinjiang-Uygur Autonomous Region is the modern name for Sinkiang Province and its forerunner, Chinese Turkestan. The capital city is Urumqi, with a population of around a million people. A thousand years ago Kashgar and Khotan were more important centres, being oases along the famous Silk Road, over which trade between China, India, and the West took place. Politically, the Chinese now control Xinjiang-Uygur Autonomous Region. The significance of the term "autonomous" escapes me.

The adventure begins

It was no simple matter to get permission from the authorities to visit the western reaches of China, as the province of Xinjiang had long been closed to foreign travel. At one stage in the preparations we almost gave up, but Russell Beck persisted, and all of us were grateful for his dogged determination, which eventually wore down our future hosts. It was part of Russell's approach to the authorities that we were prepared to offer lectures on jade by "the experts from Canada and New Zealand." I have no idea how much weight this carried, but we finally did receive our permits, at a time when Xinjiang was just opening to outsiders.

Together with Russell, the group that coalesced included Dr. Alf Poole, a cardiologist and jade fancier; the New Zealand jade artist John Edgar; Brian Ahern, a park warden in charge of protecting jade deposits; and Murray Grey, who was a bookseller with a special interest in historical aspects of jade. These five constituted the Kiwi contingent; I was the lone Canadian representative.

We visited Khotan in 1986, some 700 years after Marco Polo had passed that way. The Chinese call the city Hotien now, but somehow

Khotan has a more romantic ring to my ears. Today there are a dozen ethnic minorities in the region, the most numerous being the Muslem Uighurs. They are generally peasant farmers and small merchants.

On September 5 we stopped over in Hong Kong, where I visited Hong Kong Island and bought a zoom lens for my camera at a good price. I was the first customer of the day, and local custom dictates that the first customer must leave happy, in order for the store to have a propitious day.

In other local stores I saw some nice carvings from China: jade, some jadeite, rose quartz, ivory, lapis, malachite, and tiger-eye. One fabulous carving in jade was a metre high—all chains and ornamentation, for $238,000 (HK)! One shop had much ivory, including complete tusks. There were also camera and electronic shops all over the place. At an English bookstore there was nothing on jade, and I got the impression that few people really knew what jade was all about. They called some of the mineral jasper, which to me was nephrite. Much of the light green material was probably serpentine, although intricately carved.

My few days' stopover passed easily enough while I waited for my connection to Beijing. In the meantime, Hong Kong was a fascinating city that demanded to be explored. The Jade Market was on Kansu Street, so I went up Canton Road to the bus terminal and eventually found it. The market itself was pretty squalid, and no one seemed to speak English— quite a bit different from Pender Street in Vancouver. There were mostly small jadeite pendants and similar items, but one small, serpentine, made-in-China Buddha attracted my attention. As I walked back to the hotel past the immigration building, 10 armed officers emerged, two carrying shotguns. Later I caught a taxi out to Tai Tak airport, and two hours and 40 minutes later was in Beijing, where I was met by a tall man waving a copy of my *Jade in Canada*—Gua Lungde was to be our guide and interpreter for the duration. In the city centre at the Beijing Hotel I met up with four-fifths of the New Zealand delegation: Russell Beck, Alf Poole, John Edgar, and Murray Brown.

On September 8, before going to the ministry, we took a walk, because Alf wanted to find the Friendship House. According to the map it was only two to three blocks away. However, these were Chinese blocks. After a 40-minute walk we decided there wasn't enough time to get to our destination, so we returned to the hotel. There were bicycles everywhere, some tandem, some tricycle, and some like small trucks. The buses were crowded and mostly articulated. I did see some minibuses and taxis, but there did not appear to be any private cars. It was warm in Beijing but without the oppressive humidity of Hong Kong.

Later there was a trip to the museum at the Ministry of Geology. We drove into a small courtyard, where there were some very large mineral specimens on pedestals: quartz crystal, galena, chalcopyrite, magnetite, and nephrite. The museum was built in 1959 and is an impressive 4,500 square metres. A magnitude 7.8 earthquake in 1975 apparently affected the building. Inside we were conducted into a room for tea by director Jo Wang Why, where there was a short discussion on jade and then lots of introductions. Later there was a big lunch—10 courses of varied taste and interest but mostly seafood (I think).

The banquet director Yang Zhiling made a speech, in English, about relationships between New Zealand and Canada, then served Russell and me the first portions of each plate, as we had been seated on either side of him. Copious amounts of beer, wine, and spirits were poured. Our glasses kept filling up, and then the banquet ended with a bowl of soup and an apple—not a B.C. Macintosh, but good anyway.

The following morning we were off to a factory, which turned out to be the Beijing Jade Workshop. My subsequent notes state there were 1,800 workers, half of them female. To these people, "jade" is any beautiful stone—rose quartz, coral, and all. As was customary, we were given a speech of welcome. The factory was started in 1958. The workers were paid 85 yuan per month and included 300 designers and 7 master designers. Some 70 percent of their production was exported to over a hundred countries. One piece in the showroom was valued at $70,000 (U.S.) and took 10 years to complete.

Afterward we headed back to the hotel for lunch, which turned out to be the usual seven to eight courses. There was beer but no tea or coffee. In the afternoon we visited the Forbidden City, now called the Palace Museum, and my impression was that of spaciousness, freedom, air and serene buildings. Everywhere there were large courtyards and beautifully tiled roofs with gargoyles. At the Museum of Gems there were nephrite carvings that we examined with interest. A favourite theme seemed to be jade mountains, which were large boulders with peasants and bridges often carved on them. There were many tour groups present while we were there, aside from the Chinese and English visitors.

There are mandatory places that all visitors to Beijing ought to visit, and September 10 was a day for us to do that, as pre-arranged by our hosts. In the morning there was a long drive and rushed visit to the Summer Palace. Then, after a hasty lunch back at the hotel, we headed out to the Ming Tombs north of the city. By the time we returned that evening we felt we had thoroughly "done" Beijing. So it was with some relief that the following day was a working one. We were to present a series of lectures at the Ministry of Geology.

The lecture hall was narrow, with a double row of wooden slat benches about four people wide. Students made up most of the audience, although some were employees of the ministry. We didn't know how the translator made out—I suspect it may have been quite basic. Earlier, I had thought I should dress up a bit and had worn a shirt, tie, and jacket, but the meeting turned out to be informal, crowded, and hot, so I shed clothes. Russell Beck began with a lecture on New Zealand jade, and then John Edgar talked about carving. Thereafter I gave my presentation on "Jade in Canada." I started by bringing greetings from Canada, and it was with humility that I talked about jade to the Chinese, who were the real experts. I joked about once being an expert, saying I had published all I knew and now anyone who read my publication knew as much as I did!

Early the next morning, while in my hotel room, I discovered an English news channel on Beijing Radio on the AM band and caught up on world affairs. Later we went to the Museum of Geology, where we were first conducted into a meeting hall and given tea (of course). We adjourned to another hall for John Edgar's lecture on carving. He had good continuity and good slides on the concept of carving small pieces to wear or carry, rather than on carving mantelpiece statues. After lunch we went to the Museum of Natural History, which turned out to be pretty run down, but there was a good selection of jade artifacts. The dinosaurs were interesting, with *Tyrannosaurus* and *Diplodocus* and several others, but not all that exciting. Later we emerged into beautiful sunshine and stopped in Tiananmen Square, across from the Palace Museum (the Forbidden City), before going out to the New Zealand Embassy for a beer.

Our Kiwis were quite anxious to visit Rewi Alley, a famous countryman who became a personal friend of Mao Zedong during the Red Revolution and who, it turned out, seemed to be in pretty good health, although he would be 90 years old in December. He was regarded by the Chinese similarly to the way Canadians thought of Dr. Norman

Bethune—as a sort of folk hero from the outside who supported the Great Cause. Alley had spent a long time in China, arriving first in 1927 and starting as a factory inspector in Shanghai. He always had a strong social conscience and worked continuously, often against the official imperial policies, to better the lot of workers. In 1937, when the Japanese invaded, he founded a school to train agricultural students and was active throughout the Cultural Revolution. He now lived in a fine house not far from Tiananmen Square, surrounded by lots of books.

We were served coffee with sweet milk and a piece of tasty fruitcake, and during our visit at least three of Rewi's four adopted Chinese sons were in the house. After an hour of conversation, John gave his famous countryman a jade art piece, and Murray Brown presented a book by a New Zealand author. We all had our pictures taken as we left. Rewi Alley received the unconditional gratitude of the Chinese nation, something that was rarely conferred on foreigners, and along with Edgar Snow and Norman Bethune, formed one of the greatest foreign triumvirates. Sadly, Mr. Alley died a year after our visit, in 1987.

On September 14 we were driven to Badaling to see the Great Wall of China. En route we came to a pass in the mountains, travelled through low woods to the summit slowly gaining height, crossed a railroad track, and, finally, I saw the wall! At Badaling there were crowds of people and lots of little shops. As was traditional, we climbed the wall, walking for perhaps half a mile along the top, which was surprisingly steep in places but had steps. The handrail was clearly designed for people of a small stature and wasn't much help to large folk of European origin. I noticed that repairs were going on, with mule-loads of stone being packed up to repair damaged spots.

From Badaling we descended to the plain, skirted east around the mountains and then after a few miles turned north again to the Avenue of Animals and into a restaurant for lunch. Once again it was one of those enormous Chinese meals with a dozen delicious dishes plus beer, finishing with a piece of cake. Later we went into the garden of the Ming Tombs and for another fee entered the tomb itself, dark and hot and very crowded. No photos were allowed, so I bought a set of slides, which were better than I would have taken even if permitted.

The following morning we prepared for a noon departure for Urumqi in Mongolia. So much had happened since September 4, when I had left Vancouver, that it would be easy to forget scenes, impressions, and places were it not for keeping a diary. Eleven days were crowded with new experiences, to say nothing of new smells and sounds. And now there would be Urumqi, which would likely be truly exotic. In my diary I noted that we must get some jade pebbles, one way or another.

At the airport at one o'clock we were supposed to be loading but instead stood sweltering in the sun, waiting for something or someone. Then a whistle blew, and we all fought to get into the plane as if there were fewer seats than the number of passengers! It didn't make sense, but I have noticed the same sort of mania grip North American passengers at airports, too, to a lesser extent. Airborne, we crossed rugged, snow-capped mountains that must have approached 5,000 metres (16,000 feet), dusted with new snow, and shortly after five o'clock we touched down in Urumqi.

My impression of the city as we flew in was one of being ringed by great mountains, but once on the ground it appeared to be a flat plain, as the summits were obscured by haze. The city was neat and tidy, with many poplar-lined streets, and was not so dense as Beijing. On the one hand there were fewer bicycles, but on the other the cyclists seemed to take more risks.

We were met by a delegation that included the director from the Ministry of Geology and Mineral Resources. Despite the lateness of the hour, once in the city we called in at the Geology Museum for 30 minutes to look at their exhibits. The collection was well displayed, bright and clean, and, curiously, the minerals were labeled in English. Despite our long flight from Beijing, we visited a jade carving factory as well, where the manager, Mr. Wang, told us that 70 percent of their product was exported. Both alluvial and *in situ* jade was used. Finally, we registered at the Overseas Chinese Hotel, where we could wash up for the opening banquet, to be hosted by Vice-Director Chen Zhefu of the Bureau. In the dining hall, we had the usual 10 courses, plus beer (a light, good brew), wine, and spirits. The wine was a kind of vermouth, and the white spirit was liquid dynamite—maybe 75 percent alcohol. As was seemingly the norm, there were numerous toasts, smiles, and lots of confusion.

At 5:45 a.m. the next morning our guide phoned our rooms and told us to bring our passports and plane tickets to breakfast. I wondered why. In due course these were collected and vanished mysteriously into some official's bag. A traveller always feels rather vulnerable when that happens, for without identification or ticket, he or she is totally at the mercy of unknown officials, some of whom do not look particularly savoury. Down in the lobby it was still black outside. This was due in part to the early hour but also to the curious arrangement whereby all of China is in the same time zone. Thus those in the extreme west, such as Urumqi and Khotan, are obliged to get up at the same hour as their Beijing counterparts, although in a normal world they would be three hours behind the capital. I was grateful that Ottawa had not decreed the same sort of punishment on British Columbians.

We drove to the airport in the dark. Along the way untold road sweepers worked with witches' brooms. At the terminal, our documents were as mysteriously and inexplicably returned to us as they had been taken away. I noticed that security was a little more rigorous than usual and that security staff looked in some of the baggage. We then filed into a turboprop aircraft that was four seats wide inside, like the old DC-3s. The sign that read "No Smoking" and "Fasten Seat Belts" was in English. Just after eight o'clock the engines started, but the sun was not yet up, as it was really only 5:00 a.m. A slight ground fog made visibility poor as we taxied down the runway, but on gaining elevation, small lakes and reservoirs appeared in the brown earth. Here and there we could see long narrow fields of green that, geologically speaking, were sedimentary beds eroded to a youthful topography.

It quickly became apparent that ours was not a fast aircraft. Gua, our guide, said it would do 400 kilometres per hour, or about 240 miles per hour. Later the stewardess brought dried melon in plastic containers, which were nearly impossible to open—I kept mine for later, when we might be near a machine shop.

During the flight I took the opportunity to scan the contents of my guidebook. The ancient city of Khotan is located in the southern part of Xinjiang-Uigher Autonomous Region. The eastern part, so my book told me, is the Mongolian Autonomous Prefecture. The south is ringed by the lofty Karakash and Kunlun mountains, and the western border is Kashmir, a state fought over, and variously controlled by, both India and Pakistan. The northern part of Xinjiang vanishes into the vast expanse of the Taklamakan Desert. Like so many central Asian countries and regions, the district has enormous territory and rich resources, few of which have been developed because of a chronic shortage of capital.

The population of Khotan District, I read, was just over a million at that time, with no less than 11 nationalities, showing that the region really was the melting pot of wars and waves of immigration and occupation for many centuries. Ghengis Khan came through the area, and so did Marco Polo. Within the region are people who call themselves Uighur, Han, Hui-Tajeck, Kergez, Tibetan, Mongolian, Shibo, Russian, Manjew, and Kazuck. However, most are Uighur, who are Muslim. The guidebook advised that the people are fond of singing and dancing and have optimistic dispositions (who writes this stuff?). Further, the book said, they are a valiant and industrious people. Under the chapter describing "Geography," I learned that there are 76 villages and 11 towns, with Khotan City being the headquarters of the Communist Party and the seat of government for the Khotan District. It is also the economic and cultural centre of the region. The north and east of the country are low; the south and west of the country are high, with the loftiest mountain being 7,282 metres (22,235 feet) above sea level.

A famous local poet once described Khotan as "a rosary of oases between the boundless and indistinct desert." Less lyrical is the unhappy fact that the area has little rain. The annual precipitation is just 35 millimetres (1.4 inches), while the evaporation is 2,558 millimetres (100 inches)—you do the math. The highest temperature is 42.5⁰C, and the windy season is from April to June. Despite this somewhat intemperate climate, Khotan is said to be a good place for agriculture, forestry, and animal husbandry. Because of the limited water resources, however, all the agricultural land is irrigated. No less than 24 rivers flow down from the Karakash and Kunlun mountains, so given the annual rainfall, it's obvious that this is the main water supply for the area. The rate of flow, according to the guidebook, is 7,200 million cubic metres per second of river water, and 2,200 million cubic metres per second of underground water.

Since "liberation" by the Communist Party, all minority nationalities of Khotan have, through hard struggle, built water conservancy projects. (This sounds like a euphemism for labour camps to me.) Regardless of how it was arranged, it seemed the district had built many middle- and small-sized reservoirs and created 14,000 kilometres of irrigation canals. The end result, of course, was the development of huge irrigation projects.

Since ancient times Khotan has been celebrated for its rich farming of vegetables and succulent fruit, and its lush growth of flowers. It produces wheat, corn, barley, rice, cotton, and oil crops (not to mention marijuana). Livestock includes the indigenous Khotan sheep as well as cattle, donkeys, camels, and the small but highly regarded Yili breed of horses, famous for their resilience and surefootedness. The country is also rich in mineral resources, including coal, iron, copper, gold, manganese, tungsten, nickel, lead, phosphorus, gypsum, and mica. Oil is under production, as well as gas.

My guidebook certainly waxed lyrical about what appeared, from the air at least, to be an arid and hostile land. But it had one final gem to bestow: Khotan, I read, was world-famous for three special local products: silk, jade, and rugs. I proposed to keep an open mind and an observant eye out for those possibilities.

Two and a half hours later we touched down at Aksu for a 30-minute stop. The terrain was quite sandy, and there were lots of military personnel striding around or loitering with an official mien that seemed to say, "I dare you to attack or even photograph me." Finally, at eleven o'clock, we were off again for Khotan. Below, there were rivers and canals and plenty of green fields, attesting to the efficient use of irrigation (or the work of labourers). Unfortunately the haze was so thick that photography was impossible.

Shortly after midday the plane touched down at Khotan–Hotien, where we were met by a representative of the Ministry of Geology. Sitting in the lounge, I saw a new style of headdress: the men wore hats with no peaks. These were Uighurs. We gathered in a small room for tea and introductions to our hosts and local geologists. Here, the teacups all have lids; keeping out flies came to mind.

We collected our baggage and drove into town to the hotel, where Alf Poole and I got to share a room. The place was third-rate but clean and adequate. There was no light switch in the bathroom, but, curiously, the switch was across the hall. However, our room boasted a fan. The bed had a sort of wash basin underneath it, like an old-fashioned chamber pot, and a pair of slippers. There was also a cupboard provided, with a lock. Another thing we all noticed was the remarkable number of spittoons around the corridors and in rooms everywhere. There must have been an "anti-spit" campaign going on to cure the local bad habit.

Xinjiang was Uighur country (although the Chinese have taken over), and the people certainly dressed differently. Men wearing black waistcoats over white homespun shirts, plus baggy pants, seemed to be the unofficial uniform everywhere. In the streets there were lots of donkey carts and melons, melons, and more melons. At five o'clock it was still siesta time, and no one was stirring. I would have liked to wander about the town, see what was for sale and take some pictures, but I didn't.

At six o'clock there was a meeting in a big room with fans going and geological maps spread out. We received a warm welcome from the boss of the Geological Brigade, who told us something of the Khotan jade business. Here the jade is formed from dolomitic marble. There are at least 10 locales, and they claim various kinds: the colour varies from white to yellow-green, green and white, white with black spots, and black. We listened to the lecture, but the translation made it difficult to follow at times, and my notes were sketchy. I did gather that there were many jade deposits *in situ* and that the Brigade geologists classify valuable deposits as:

1. Precambrian dolomitic marble–granite contact— both green and white jade.
2. Late Precambrian deposits in the Kunlun Mountains— black jade in the Karakash, and white jade.

Alluvial jade in boulders up to several tons was scarce. Most of the cobbles weighed between a few kilos and a few hundred kilos. We also learned that in the Kunlun Mountains the weather can be bad and transportation difficult. For instance, a famous black jade locality was described as being "a 10-day donkey trip." Some jade had been packed out from that locality, but the site was at the elevation of 5,000 to 5,700 metres (16,500 to 18,500 feet), and partly glacier-covered. These were Quaternary deposits.

According to our translator the moraines were "full of jade," which sounded promising. The talk was upbeat, although I was somewhat disappointed to have come all this way for what looked like very little. I am a geologist and so was interested in *in situ* deposits. However, these all seemed to be above the 5,000-metre mark and were not included in our itinerary. This was somewhat surprising, as there are accounts of *in situ* deposits at lower elevations along the upper reaches of the Karakash River. More likely it was the military sensitivity along the border, rather than the altitude, which precluded our visiting the areas. The region just to the south of the mountains has been disputed by India and China for decades, and the Pakistan army was also active on the western flank.

After breakfast on September 17 we were off for a look along the

Urungkash River, which was wide, fast-flowing, and milky. We walked along the boulder fields on the banks and saw granite, micaceous schist, serpentine, and red feldspar porphyry. Not surprisingly, jade was scarce. July and August were the high-water times, after which the locals hunted for jade, so now was the right season. The entire area was very dusty and there wasn't a blade of grass anywhere—just sand, pebbly gravel, and boulders. We hunted until two o'clock, when we stopped for lunch. Russ, John, and Brian found a nice piece each, but I was skunked, and Gua, who was with me most of the morning, found a poorish boulder, too big to bring back. However, I broke off a few pieces, which said something for the quality.

After lunch I walked upstream to a gorge, but there appeared to be no jade. On the way back I met some soldiers along the road. They were not armed, so I gave them a salute, to which there was no big response, although some showed a sense of humour. As the afternoon progressed, the wind blew hot, and little dust storms picked up. Russell came back with a 40-kilogram boulder, over which Brian said a Maori prayer; although Brian isn't actually Maori himself, the prayer is a custom. The

New Zealand carver Brian Ahern (left) says a traditional Maori prayer over a newly found jade boulder next to the Karakash River in Xinjiang Province, China.

boulder was not a prize in my eyes, but both Kiwis thought it was great.

On the way back to the hotel we visited a carpet factory—remember: silk, jade, and rugs—where we had tea and melons in the showroom before entering the factory. Three people sat on a makeshift bench in front of a loom, weaving and hand-tying. About a dozen double-sided carpets were being worked on, mainly by women, with one child and a few older men. There was no light in the building; the only illumination came from a skylight. It looked somewhat primitive to me, but the carpets were probably correspondingly cheap.

The Uighurs are of Turkic stock, short and dark, but not noticeably darker than we are with a tan. Their main transportation seemed to be donkey carts, and the locals appeared a pleasant people, with ready smiles and a general willingness to be photographed. Some of the older men wore a Tatar type of hat, like the Cossacks. For dinner there was plenty of beer, tea, fish and chips, mutton, hard-boiled eggs, plus some unknowns. There was also brown fried rice, which I covered with sugar and called dessert—all in all, a pretty good meal.

The next day after a breakfast of more hard-boiled eggs, tea, dumplings and jam, and peanuts, we loaded up for what turned out to be a 200-kilometre trip. There had been no rain in the region for many months, and some of the vehicles had no windshield wiper blades. Our destination was the famous Karakash ("Black Jade") River, but first there was a stop at Moyu, which is on the Old Silk Road. Thereafter we arrived at a placer mining area, where six terraces had been built along the river, covering some 20 square kilometres. We were told that diamonds, gold, and jade had been found in this terrane. The gravel was said to be 50 metres thick, and it seemed anyone could dig there, but presumably no great fortunes were made. A few prospectors were present as we watched, and a local geologist explained that kimberlite bedrock had given rise to the alluvial diamonds, but there didn't seem to be any interest at the government level to develop the properties.

This was a real desert, with, again, not a blade of grass or vegetation of any kind. In due course we turned down off a terrace on the way back to the Karakash River and stopped at a local house for refreshments. It was nice and cool inside, and we sat on rugs on the floor. The occupants were friendly and brought in peaches and grapes, so we tucked in with relish while the family looked on. The room was largely unfurnished. Clothes

Xinjiang Province: Hunting for black jade cobbles on the banks of the Karakash River, which flows north out of the Karakoram Mountains into extremely dry conditions on the edge of the Taklamakan Desert. From left to right: geologist-host Mr. Zhang, translator Gua Lungde, Stan Leaming and Ms. Han.

were hung up to dry, and in another room a baby cried. Soon we were on our way again, but we stopped to cool the engines, which seemed to be overheating (a perennial problem in Asia). Later we saw the Karakash River again, down in the valley bottom, and our geologist host Mr. Zhang gave a short lecture about the late Triassic beds we could see in front of us.

At 3:15 p.m. we stopped to water the jeeps again at a spot where the road cut through a strikingly formed rock arch of Carboniferous conglomerate. In a wet climate a feature like that wouldn't last more than a few centuries, but here in the rain shadow of the Kunlun and Pamir Mountains it's likely only the wind that erodes it. An hour later our lunch was about finished, and we packed up and set off upstream for a few kilometres, on another jade hunt by the river. Young Uighur children came along, so I took some photographs. We didn't have much luck with the jade prospecting, however. In the heat of the afternoon we chose instead to visit a local farmer for grapes and watermelons, eating them in the cool of the shade by his building.

We reluctantly left our comfortable refuge to return to the road leading to the Karakash River, and reach the river we eventually did. From that thin, wet ribbon in an otherwise drought-stricken land, the barren hills led up to great summits of rock, often cloud-covered. Distant glaciers, momentarily glimpsed, added to the land's contrasts. The sheer scale of the region, its emptiness and its remoteness all combined to make us feel displaced, disconnected and as though we were somehow in a dream, where nothing made sense, yet nothing had to make sense. It was simply there, all about us, in gigantic form—a backdrop for a stage where there seemed, to my mind at least, to be no play waiting in the wings.

At the next collection of huts the villagers came out with their jade boulders, and I bought three pieces, all pretty cheap. I suspected the sellers were happy too, as they were cutting out a lot of middlemen and getting a better price than usual. So everyone was pleased with our stopover. It was late afternoon as we set out for the hotel, and on the way it became pitch dark. The driver careened down the road, missing pedestrians and donkey carts by inches. We somehow arrived safely and had dinner after scraping most of the day's dust off ourselves.

I was awake at 8:30 a.m. (standard China time), but it was another hour until breakfast. There was no hot water in the bathroom, and on looking outside it was barely light, with the streetlights still on. Loudspeakers were making strange noises, likely *muezzins* calling the faithful to morning prayers. We had a thermos of hot water with us in the room, and after Alf Poole sterilized it, we made a cup of coffee. Once down in the courtyard, we watched merchants set up their displays of jade on rugs, but they weren't doing much business that early.

Some time after ten o'clock we headed out on another jade hunt with Mr. Zhang as team leader and principal geologist and Mr. Abu-du as local Uighur translator. En route we again stopped for watermelons, which was fast becoming a tradition. At the site we mineral-hunted for awhile, not finding much, but, nevertheless, we finished a late picnic lunch with spirited toasts of beer. A great crowd of locals had gathered around us, as was so often the case in Asia, where no one, it seemed, had anything better to do than stand and watch a group of strange "round-eyes." It was good to know we provided some entertainment value.

At five o'clock that evening it was our turn to be entertained as we pulled into a village for a concert. A five-piece orchestra and some singers were on the stage. On the floor there was a grape arbour with electric fans and chesterfields arranged around the sides of the room. Three girls danced a kind of step dance that resembled a Highland fling. There was waving of arms and twirling about in green, white, and lavender costumes, and then along came a load of watermelons, which was certainly a change from the popcorn eaten in North America.

The dance was followed by another announcement in Chinese, which triggered four male dancers to join a girl singer for some vigorous dancing in the style of the Cossacks. This was followed by another act—a girl in a beautiful costume leading the orchestra in a quick-step; a drum and fife and mandolin-like instrument joined in. Then a drum and fife and mandolin-like instrument. The girl was young and pretty, in a long green dress with a matching vest. The tempo picked up and the orchestra jammed in. Round and round spun the girl, with her pigtails a-flying, all to much applause. Other acts followed. Then a Japanese tourist joined in, and the girl tried to get more of the audience into the act. A Japanese woman got up too, and things began to get busy. Brian went onstage, as did Russ, who did a Maori dance—at least I think that's what it was— that kept everyone awestruck. Finally the concert was over and we left for the hotel, getting back in time for supper. Russ said he was thinking of changing careers, and going "on the boards" full-time.

September 20 was a lecture day, so I shaved and put on clean clothes. Through my window it looked like rain, which was likely an illusion because of the haze, and from outside I could hear the loudspeakers in the square blaring as always (or so it seemed). We assembled for the talks at the geology department, where about 40 people were present, including two women, which was nice to see in that male-dominated society. Tea was served in mugs with those little lids again. The room was, unfortunately, not equipped with blinds, so it was somewhat unsatisfactory for showing slides.

The proceedings got underway with a short speech by the director of the 10th Geological Brigade, and then Russell Beck made his presentation. He had words of praise for the hosts and expressed the group's gratitude for the opportunity to see the jade fields of Xinjiang. He reflected on the fact that the Maoris came from China via the Pacific islands a thousand years ago, found nephrite, and formed adzes and tools from it with sandstone saws. No metal was used. They also made neck ornaments called *hei-tikis*, which are small human-like forms. They fashioned clubs up to 40 cm (15 inches) long for weapons, as well as jade tools for woodworking.

He continued with a summary of the geology of New Zealand, jade formation, and the areas where the stone is found. Russ told of gold and jade in the alluvial gravels along the west coast of New Zealand's South Island. Dredges were first used for extraction, and at one time there were 40 of them in operation. His talk was illustrated by slides, but there was, curiously, only one tray available for use, so we had to pause to reload. In New Zealand, unfinished jade cannot be exported, and there are special jade reserves in some of the national parks. It was a good talk, and I learned a great deal. I suspect our hosts did too.

After a break I gave my talk and showed slides on the Canadian jade scene. Afterward there were questions, some of which were difficult to answer, and there was a problem hearing the softer-spoken students. Finally, at 2:00 p.m., we broke. There was to be a banquet at 7:00 p.m., so Alf made a quick trip out to the carpet factory by bus, came back with some rugs and gave one to me. Later we walked to the post office, where we discovered that the stamps were not pre-glued; one had to use a glue pot.

Still later we visited an historic site where Aurel Stein is said to have dug, and our local translator, Abu-du Kirim, told us that lots of pottery could be found. I picked up a fragment for a souvenir. The old city there (dating back to 800 AD) is covered by three to nine metres of sand now, as it's on an alluvial fan of the Urungkash River, which has moved over five kilometres during the intervening millennium.

Back in the jeep we still had time to spare, so we headed off to visit a jade factory, stopping along the way for the inevitable watermelons. At the factory we were given tea and an introduction that was something of a repeat from the Geology Brigade: white jade is from the Kunlun Mountains at 5,000 to 5,700 metres above sea level. As well, snow occurs all year round, so it's not suitable for mining; most of the jade that is recovered has been glacially transported to lower levels.

In 1972, when U.S.–Chinese relations began to thaw after the Cold War, U.S. President Nixon was presented with a white jade carving, which he had requested. The Urungkash is the White Jade River, and a yellow-skinned white jade is found there. Even around the city, it is said that jade has been found after the annual spring flood, but production is small; estimates are just a few kilograms annually.

Green jade from the Urungkash is rare, and black jade comes mainly from the Karakash River. Occasionally the black and white occur together. In the factory shop I bought two jade pebbles. Interestingly, outside the factory at that time, a truck was unloading jade boulders, some as large as 50 to 500 kilograms.

We arrived back at the hotel in time for the banquet, and as we filed into the foyer for introductions, we were each given presents: Khotan jade and a piece of silk. There were 24 guests, and I noticed that one table was curtained off. I suggested the barrier be removed so the "help" could join in. There may have been some reluctance, as these were the drivers—mainly Uighurs—and of lower rank. However, we were in the "people's" egalitarian Republic of China, and equality prevailed. I think the Uighurs were delighted.

There followed innumerable speeches and replies, plus many courses comprising duck, chicken, beef, fish, fruit, vegetables, soup, and more fruit. Throughout all this, various toasts were delivered, including one by the director. This was followed by a few impromptu acts, mostly by the westerners, but a couple by the Uighurs. As always, there was lots of wine, spirits, and beer. We toasted Marco Polo, Confucius, Rewi Alley, Bethune, and pretty well anyone else that any of us could remember, all to great applause. Then there were further songs and dances, and so the evening ended.

Later I had a shower of sorts; the water from the hot-water tap was not quite as cold as it was from the cold-water tap. It had been a great evening. Alf confided to me that he thought the gift of silk he and I received was due to our host's embarrassment at the flooding of our toilet and the need to move us to another room.

It was Sunday, September 21, early morning, and there were dogs barking and a call to prayers, but the noise level was generally low, with no music from the loudspeaker in the square, for which we were grateful. There were some jade peddlers outside the restaurant, and I bought a pebble for two yuan (about 50 cents), while Alf Poole acquired another couple for three yuan. Over breakfast we compared our purchases, and as a result I bargained for a couple more afterward.

The day had been set aside for an expedition to the edge of the desert to ride camels. Actually it turned out to be a tourist misrepresentation, as we ended up sitting momentarily on camels, taking photographs of each other. The thought of travelling any distance on one of those animals puts me in mind of Oscar Wilde's famous quotation about a horse: "dangerous at both ends and uncomfortable in the middle."

We stopped at a cluster of mud huts, and within minutes about 40 children were following us. For so few houses there seemed to be an inordinate number of kids. However, Abu-du explained that there were generally six to nine children in a family, and large households were common. It was a nice, warm day, as always, with the usual few flies buzzing around, although I'd noticed that flies were quite abundant in the villages, especially around food displays. There were also a few pigeons and an occasional raven overhead, and Abu-du claimed there were ducks in winter. We had an inevitable watermelon stop at the edge of the desert beside an irrigation canal with fish ponds before returning to the hotel to clean up for lunch. Then out to the bazaar in the afternoon, where a variety of merchandise was offered every Sunday.

The crowd was by and large good-natured, but there was nothing of interest to tourists: yard goods of cotton and silk; headwear of all styles; ball-bearing races for vehicles and carts; shovels and other farming tools; a street full of home-cooking appliances such as pots, pans, ladles, and urns; a locksmith; clothes aplenty; and melons. Lots of melons. There was also a great mass of people, and everywhere we stopped to take pictures, a throng pressed in. I felt claustrophobic. Women nursed their babies, and hawkers cried out their wares. A plethora of donkey carts, bicycles, and an occasional motorcycle honking through the crowd filled the streets. It was a fascinating scene but difficult to photograph, as there was no high vantage point.

The bazaar was huge, with many narrow side streets off the main thoroughfare. Everywhere workmen either repaired bicycles or built things right on the ground. I didn't see a single proper workbench, yet the ingenuity of the craftsmen was amazing. In my diary that evening I wrote:

Tomorrow we depart for Urumqi, so it's probably a good time to recapitulate. In retrospect, the hotel was substandard—there were few services, a key girl, and hand laundry. So-called 'hot' water came on just once a day in the evening. Everything was dusty, no doubt because of the lack of rain.

In general terms, the weather is hot, but the humidity is low, unlike Hong Kong. Food is typically Chinese, except for the French fries. It appears there is a universal cuisine after all. The hard-boiled eggs have not been encountered elsewhere on our travels. There's also a type of straight donut that is unusual.

The trips up the Urungkash and Karakash rivers were fabulous, even though we did not find much jade. However, we were able to buy some nice specimens at cheap prices from the street vendors. The region offers wonderful scenery, although where not irrigated, the land is a desert. People seem to be everywhere, and live in mud huts. However it rarely rains, so this is not as bad as it might otherwise appear. Donkey carts are ubiquitous. On the roads we came across some fierce-looking men, but the women are by and large very attractive and colourfully dressed. The Uighur concert was certainly memorable.

That evening we all went to the Hotien Theatre for the evening's entertainment. There was a live concert with some fine artists, and it lasted from ten o'clock until midnight.

On September 22 we were getting ready to leave. In the half-hour before breakfast I sat outside the hotel as the sun tried to shine through the perpetual haze—perpetual since we arrived, at least. Loudspeakers were blaring in the square and echoing around the buildings. People stood about, some came and went, drivers washed their vehicles. What a beautiful place this would be, I thought, with a little more grass! Everything grows well with all the sunshine, but it rarely rains, so water must be added. Yet there didn't seem to be a water shortage. The two rivers are an enormous supply, and they flow into the desert to the north, either to evaporate or contribute to the groundwater.

My impressions of Khotan will remain vivid: the Kunlun Mountains, views along the Urungkash and Karakash rivers, the donkey carts, the fur hats, a camel "ride" in the desert, and melon stops along poplar-lined roads. There were our horn-honking drivers, the crowds at the bazaars (and everywhere else we stopped), and marijuana plants in the cornfields. There was no doubt that our hosts had gone to great lengths to make our trip interesting, instructive, and memorable.

Then there was Abu-du, employee of the Department of Foreign Affairs and quite a linguist—or he will be. Abu-du was a young Uighur who spoke Chinese as well as his own language, and his English was quite good. We didn't discover this until we were about to leave. He did not have as rich a vocabulary as Gua, but his accent was better. I think he could converse in Russian too. Soft-spoken, he was always able to find nice, cool melons on our outings. He sometimes appalled Alf with his unsanitary methods, mind you—like washing the melons in the canal, not exactly the purest of waters.

He wrote very neatly and kindly loaned me his notes, in which he expressed himself surprisingly well in English. Regarding Khotan rugs he wrote:

> Khotan also one of the native places with the east style rugs. Since the long period take Khotan as the name of Xinjiang rugs were enjoy great reputation in abroad. Piece of remnant rug at the time of East Han dynasty was excavated from Niya, ancient city in autumn 1857.
>
> Up to now it is the oldest rug which was found by our country. Khotan rug take the famous Khotan sheep with different quality and coarse wool as the raw material. Well elasticity, good springy and soft handle and plentiful and well pressed. After stop underfoot wool can be restore at once. Khotan rugs new and original pattern, beautiful colour, simple and elegant.

On the way to breakfast that day the jade sellers were on the job in the square. There was a nice, white piece for which one of them was asking 200 yuan. We scoffed at this exorbitant price and offered 10 yuan. He refused, and we went in to breakfast. Shortly afterward Abu-du came in to say he'd talked the owner down to 30 yuan. Russell bought it. Then Abu-du returned with three pebbles for 10 yuan, so I bought those and now had some gifts. Remembering the Khotan guidebook, I reminded the others that the region was known for silk, jade, and rugs. We had covered only two of the three, so after breakfast we made a quick trip to a silk factory outside town, where we learned there were 1,650 workers. We saw the process, and bought a couple of cocoons, but were in too much of a hurry to do the visit justice. I also bought some silk fabric in the showroom before we hurried back to town.

By noon we were at the airport, and after saying goodbye to our hosts, we left for Urumqi. The Khotan Geology Brigade waved us farewell from the terminal building. I had my camera and took a few shots out of the window, but it was still very hazy. When the stewardess appeared I anticipated something to eat, but no, it was a gift, labelled in English, "Advanced Tie Clip." Alf Poole put his on his ear and got a big laugh. Later we were served tea in plastic cups with no handles. It was very hot, but a good variety. I noticed ice around the aircraft window, which melted such that the water ran down the cold wall beside me—not something I'd seen in Boeing products. There was not a lot of condensation, but it seemed unsatisfactory.

An hour later we were still flying over desert. We guessed we were heading for Aksu, where we had stopped on the way in. The captain, a happy-looking guy, appeared at the cockpit door and wandered aft. He had four bars on his shirtsleeves, so I *assumed* he was the captain. The plane roared on without him. Then the first officer came out and went to the toilet. I assumed *he* was the first officer, because he had three bars on his shirtsleeves. By then I was hoping that someone was minding the store up front. Later the stewardess passed around dried melons in childproof plastic containers that caused the big people trouble too.

Around two o'clock we landed at Aksu, amid irrigated green fields and tree-lined roads. Conditions in the airport restaurant didn't look too sanitary, so I chose to fast. We reboarded, and from somewhere, hard-boiled eggs materialized, and they were good. I wondered if the stewardess could find some more handouts, as we still had a couple of hours flying ahead of us over the Tarim Depression. By mid-afternoon, along came boxes of mandarin-orange juice, labelled in English and Chinese: "Best before 87-5-27." It was currently 86-9-22, so I was reassured. Below the plane, it was still very hazy, so to pass the time, I made a list of the things I had bought or had been given while in China to date and figured out that the silk purchased in Khotan had cost me $23.40 for 10 metres.

At 4:30 p.m. the plane's wheels were lowered, and we made our final approach to the airport at Urumqi. Outside the chilly interior, the temperature soared to 30°C. We were met by people from the Ministry of Geology, some of whom we'd encountered before, and our passports were collected by airport staff. On the way into town I noticed fall colours on the trees. Later, at the Overseas Chinese Hotel, where Russ Beck and I had a suite, there was some confusion over baggage and room numbers, but it was all finally sorted out, and the view from our room was wonderful. Dinner involved lots of beer, which we drank with relish.

That night I finally had a good bath and shampoo, using lots of hot water, which up to now had been something of a luxury. There was even a TV in the room, but all I could find was a tennis match and some car races. The only alternative was a channel with a weather map and good graphics of temperatures in various cities. The doorbell rang unexpectedly, and in came a load of grapes, pears, and melon slices.

The morning of September 23 arrived, and we'd all had a good night's sleep. As a bonus we woke without a morning cacophony of loudspeakers, *muezzins*, and dogs, although at seven o'clock, it was still dark outside. A big thermos of hot water stood in the room, with a push pump to dispense the contents, so I tried a cup of tea. But the results were poor, as boiling water works best.

The day's plans were to go to Tian Chi (Heaven Lake), so named for a girl from heaven who came to wash her feet there. Passing through the city, which had a lot of construction activity, we reached the open plains, where camels were grazing and the fall colours were in full splendour. By and by we came to a "Welcome to Tian Chi" sign in English.

We toured the lake by boat, while around us were mountains of about 5,000 metres with snowy summits, well treed lower down with both coniferous and deciduous species. Leaves were turning colour in some places. This was a glacially formed lake in volcanic rocks of Carboniferous age. At one stage we landed on the rocks and took some pictures, but sadly there was a great disregard for the environment, with much litter everywhere. The time on the lake was really fine, though, with the air cool and the day warm enough for comfort. The lake was at 1,900 metres, which may have explained why I was puffing up the hill to the parking lot. Back at the hotel a good soak was welcome before supper, although the laundered clothes that I'd planned on wearing came back wet, so I had to hang them up in the bathroom under a heat lamp. A string clothesline was provided in the room, so I suspected that many guests did their own laundry.

Local expert Mr. Yang Han Chen had, the previous year, co-written a book on Xinjiang's gems and jades, and at dinner that night we were each given a copy, but they were unfortunately in Chinese. Later I had a meeting with him and Mr. Gau in an attempt to establish an import business in minerals. They had a wide variety of products, were eager to export and no doubt wished to earn foreign-exchange credits. I was given a beryl crystal, but nothing resulted from our meeting. I later learned that an English version of Mr. Yang's book was to be published shortly, and by keeping in touch with the printer in Hong Kong I subsequently succeeded in getting a copy. However, there were still bits

I couldn't decipher. The text was plagued by typos, misspellings, and muddled grammar and had no doubt been put together in a rush. In spite of the book's shortcomings there was much to learn from it. I was surprised to discover there were 44 different kinds of jade, not two, as we generally acknowledge in the West (nephrite and jadeite).

We were up very early on September 24 to lecture again. At eleven o'clock we started the presentations in the hall of Mr. Yang's laboratory building. About 50 people attended—mainly men but a few women. I was embarrassed after my presentation, as my audience knew as much as I did, but at least I could tell them something about Canadian jade occurrences. Russell lectured for an hour on the New Zealand jade fields, and once again, the room wasn't dark enough, so the slides were poorly visible. In addition, the projector acted up. On a positive note, though, there were two slide trays available at this venue.

Thereafter John Edgar gave a talk on modern jade carving in New Zealand, explaining the tools and techniques. Later we had lunch and then visited the Museum of History, where there were some good exhibits of minority dwellings, such as tents in the styles of the Uighur, Mongol, and Tatar, as well as mounted animal exhibits, including sheep, goats, and horses.

That evening we began to pack up in preparation for travelling again the next day. After supper I headed up to my room, but inexplicably, the elevator stopped and the door wouldn't open. I was not alone, however, since all elevators are operated by a member of staff. In this case it was a young lady who seemed somewhat nervous at being stuck in a tiny cubicle with a round-eyed barbarian, even if he was over 60. In due course she managed to reach someone by phone, and then we waited. In the meantime I had a lesson in Chinese. An hour later the elevator restarted, and I belatedly reached my room.

The following morning we were all up before dawn for our flight. With breakfast over, we took a taxi to the airport, where hundreds of passengers milled around. All announcements were in both Chinese and English, which helped us get our baggage checked through security without problems; we arrived in the loading lounge just as dawn was breaking. Although my mind was still in neutral at that hour of the morning, I noticed that the toilet signs were marked by heads—males with a cap and females with braids, and half the fluorescent light fixtures were out.

We finally boarded a TU-154, which is a three-engine, Russian-built jet similar to a Lockheed 1011. As we were walking up the ramp, two biplanes flew over, looking like something out of the First World War. On board, an announcement in both Chinese and English told us we were due in Shanghai in four hours and that the plane would fly a speed of 950 kilometres per hour, which was somewhat faster than our recent air transport. We were again in the present, after a fascinating trip back through time, space, and history.

1987: Taiwan—Rekindling an ancient passion

The island of Taiwan off the east coast of China has been variously known as Formosa ("Beautiful Island"), the Republic of China—not to be confused with the People's Republic of China, which is the mainland—and more latterly as Taiwan. The Chinese first arrived in 607 AD. In 1602 the Dutch occupied some of the island, as did the Spanish later. The Manchus cleared the island of foreigners in 1662, but it continued to change hands, being a Japanese possession from 1895 until 1945, when it reverted to China. In 1949 the retreating Nationalists under General Chiang Kai-shek withdrew there from Mao's advancing Red Army on mainland China, and an uneasy truce has prevailed between the two Chinas since.

Whatever the politics of the country and its name, in 1986 it was still not possible to fly directly from China to Taiwan. Instead we travelled via Hong Kong (then a British enclave) to Taipei, the capital, where Russell had arranged a visit to the jade mines. Up to the early 1980s, some 50 percent of the world's jade production was claimed to have come from the Fengtien region, a small geological area about halfway down the east side of this 14,000 square mile island, where talc is also mined in considerable amounts. However, from 1975 production had declined to the point where, a decade later, there was almost no output at all. E.K. Carpenter wrote that in 1968, when he visited the mines by road, the route was marked by sharp bends and curves, so that "meeting an oncoming vehicle (was) earth shattering."[2] No doubt there was a story behind that observation. The central spine of the island comprises peaks up to 4,500 metres (14,000 feet) and is extremely rugged in places.

Our first glimpse of the island showed green fields, lush valleys, and highways. The airport (named in honour of Chiang Kai-shek) was impressive, and we drove the four-lane (paved!) highway to Taipei in style. After our previous month of dust it made a pleasant change. Our guide was to be Professor Tan Li-ping from the geology department of the National Taiwan University. As Dr. Tan had studied in North America, it was a pleasure to dispense with the necessity of an interpreter.

The first order of business was to visit a jade shop, where we met Professor Na Chih-Liang, an expert in Chinese jade and at one time a deputy director of the Palace Museum in Beijing. Dr. Na showed us his two-part book on Chinese jade art, but it was unfortunately in Chinese only, which rather limited its usefulness to us. Nevertheless, Russell bought a copy, and I later regretted that I had not. The store displayed the works of contemporary artists, some of which were quite beautiful pieces, with correspondingly high prices. One of the island's better-known sculptors was Tu Li-Kuen, who had a piece on show valued at the equivalent of $64,000 (Canadian).

The next day we avoided E.K. Carpenter's earlier experience by catching a train for Hualien on the east coast. I learned that tickets were difficult to obtain and that Dr. Tan had been quite creative in getting them. The train curved toward the mountains shortly into the trip, and we were treated to countless tunnels and bridges. The track passed numerous towns and villages along the Neelung Ho, a placer-gold river from which more than 100 tons of gold had been officially extracted, but which Professor Tan estimated was more likely to have been 500 to 800 tons in the previous century. In the early 1600s a party of Spanish or Portuguese (it was never confirmed which), heard about the river, but their expedition proved costly, as 100 headless bodies were later found in 1662.

In due course the train emerged on the east coast, where every river delta was intensively covered in rice and corn fields. At one point I spotted a curious sign which read "Tungshan Lions Club" in English. The coastal plain began to widen, with an accompanying increase in harbours, fishing boats, palm trees, and inland, paddy fields. The tracks crossed a number of substantial rivers, and to the left, attractive beaches could be seen. The island is quite beautiful. Finally, a few more long tunnels, and we arrived at Hualien, where we were met by the secretary of the Hualien Tourist Association and escorted to our hotel, which had round windows. I wondered about hobbits but saw none. Below and across the street a neon sign proclaimed ICBC—not the Insurance Corporation of British Columbia, I hoped.

The itinerary included a visit to a marble works, where massive saws were slabbing marble for tabletops and counters. There were also numerous rock lathes at work, turning out vases of varying sizes and shapes. From there we visited the Toroko Gorge, a popular tourist destination that was spectacular and deserving of its acclaim. Back at the hotel we prepared for a banquet hosted by Mr. C.C. Liang, owner of the inscrutably named Ideal Mining Company. It turned out that Mr. Liang was a man of considerable consequence, being a well-known entrepreneur who owned fast food chains and substantial construction businesses.

The hotel was really first class, and many of the staff spoke English, or at least understood it, and that made communication very much easier. After dinner we attended a concert put on by the local ethnic minority, the Ami. The performance took place on a circular stage, surrounded by the audience, and was most entertaining.

On September 30 we left for the jade mines. The road there had been badly damaged during a recent typhoon, and the journey was expected to be slow. Actually, we learned that the road had been quickly repaired in order to allow us to make the trip. Professor Tan also informed us that the current financial situation in Taiwan was not good. High inflation and high workers' wages had priced the Island's manufacturing sector out of the very competitive East Asian markets. This was true of the jade industry too. On the drive up we stopped periodically so that Professor Tan could explain the geology of the region.

The local Fengtien ultramafics have been completely changed into serpentinites and occur in a black schist unit along with marble. Asbestos, jade, and talc are associated with the contact zones. Since the structure sits close to Taiwan's east coast, which is a major tectonic plate boundary (the Asiatic Plate to the west and the Philippine Plate to the east), the region is characterized by heavy folding and shearing and by recurrent earthquakes, of course. The jade lenses tend to be small but numerous. There were reported to be a large number of mineshafts, or adits—many accessing more than a single deposit.

Production ranged from a few kilograms to several hundred tons. Once brought to the surface, the jade is generally carved by local family-run factories, which have a tradition of excellence; the families take pride in their workmanship. While some of the jade is turned into carvings, a large proportion of it is still being made into oval cabochons for jewellery settings.

Our first stop was at a talc and serpentine mine and afterward at the S-3 adit of the Ideal Mining Company, at about 600 metres (2,000 feet) above sea level. There are many local contacts between the serpentine and schist, and the S-3 was considered a good example of the structures. Typically they are lens-like beds up to two metres in depth and about 50 metres long at the most. The underground work was being done by blasting, which resulted in a lot of jade damage and loss. According to Professor Tan almost 90 percent of the nephrite was consigned to the dumps because of this short-sighted method of extraction, and this no doubt explained why the Taiwanese businesses, when presented with Canadian jade supplies in 1970, switched to that supply source.

It appeared that the nephrite reserves in the Fengtien area were substantial. Dr. Tan estimated them to be in the hundreds of thousands of tons, but the cost of mining underground was prohibitive. Most of the jade was being processed in the more than 600 small factories on the island, with the result that upward of 200,000 people were directly or indirectly involved in the jade manufacturing business. During the last war Taiwan was controlled by the Japanese, who had an active mining program for asbestos. Along with the rock fibre, they stockpiled jade, but at the end of the war, and with the subsequent arrival of the Chinese Nationalists under Chiang Kai-shek, nobody seemed to know what happened to the stockpiles.

At the S-3 we had a chance to look around and later collect whatever we could find at the portal or on the dumps. Mr. Chiang then took us to one of his manufacturing operations, where the specialty seemed to be bracelets by the bucketful. Long strings of them hung around the shop, some priced at over $50 (Canadian) while others that appeared almost the same were just a dollar. There being no apparent difference, I bought some lower cost items, which I thought would make nice presents, and we all left well-satisfied.

Behind the shop were piles of jade blocks amounting to several hundred tons. We were told they were from the Hualien Mines, but in all honesty I could not detect any difference between that material and Canadian jade. In view of the large exports of nephrite from Canada, it was possible there was some B.C. material there too. We then enjoyed a seafood lunch before catching the train back to the capital. I had forgotten just how comfortable it is to travel by train—no security, x-rays, policemen, baggage checks or waiting, and you can get up and walk around at will.

The following morning we were taken to the Central Geological Survey, where we repeated our now oft-presented lectures on New Zealand and Canadian jades. Since nearly everyone spoke or understood English, we dispensed with interpreters, and everything moved along very smoothly. Afterward I met Christopher Fong, a Taiwanese national who had spent a decade with the Newfoundland Geological Survey. As a result I learned a few new Newfie jokes. Later we visited the National Taiwan University to see some ancient jade artifacts, and this was followed by a visit to a private collection owned by Mr. Ying Chou Hsu, all of which was most interesting.

After freshening up at the hotel we were taken to Mr. Liang's private club. (He had hosted us in Hualien as well). The club was a very grand and exclusive place that had, we were told in hushed whispers, only 300 members. We spent some time at the bar waiting for our host to appear, and when he did, we were shown around the facilities, which included a swimming pool, bowling alley, pool rooms, and several restaurants, all decorated very elegantly. Mr. Liang ordered a bottle of V.O. Hennessey cognac, followed by another, and we proceeded to drink innumerable toasts to every conceivable cause. This was followed by a grand dining experience in which the food was really first class. Mr. Liang brought along his male secretary, who appeared to be more of a bodyguard and who also acted as his chauffeur. Well into the evening Mr. Liang explained that his secretary was also a talented classical guitarist, but I regret we did not get the opportunity to appraise his ability.

Another museum day on October 2: this time it was the Palace Museum, where much of the loot brought out by Chiang Kai-shek in 1948 is kept. Apparently only a small fraction of the total is on display at any one time. There were four floors, each dedicated to a particular aspect of Chinese culture. We were, naturally, most interested in the jade displays, which, like everything, were well presented, well lit, and had both Chinese and English labels, which was thoughtful.

The Palace was crowded with schoolchildren, locals, and foreign tourists, and, set against a wooded hill, the building was quite beautiful. Photography was strictly forbidden inside, so I bought some slides that were mainly concerned with jade. Late that afternoon Mr. Liang drove us to the airport, where we thanked him profusely for his time and hospitality. He had made our stay in his country very memorable. And so, after saying our farewells, we boarded a plane for the last leg of our journey, Korea, to visit that country's jade mines.

1987: Korea — Between two giants

The country of Korea has always known that it was a "nut" between the large nutcracker jaws of China and the Soviet Union and has learned to live with this situation. In few countries today have citizens managed to hang on to their traditional lifestyle while embracing Western capitalism so completely. Older women still wear the *han bok*, a traditional flowing dress, and old and new fashion are both evident in shop displays. In times past, Korea provided jade to China in considerable quantities, obtained mainly from *in situ* deposits that were mined. This was in contrast to the Chinese sources, which were scavenged from the rivers of Xinjiang Province in the far west. Over the years, however, the Korean sources have become depleted, and in the late 1970s there was just one mine still operating in South Korea, near the city variously spelled as Chun-Chon or Chanchong.[3]

On October 3 we arrived in Seoul, the capital, and took the airport bus downtown, where we were deposited near our hotel. This presented a bit of difficulty in actually finding the place, given the language barrier, but a couple of helpful Koreans with a smattering of English came to our aid, and in due course we found it. One of the Koreans, announcing that he was a Christian (crosses abound throughout the city), was kind enough to help me with my rather heavy bag up to the fourth floor of the hotel, where I arrived exhausted and hot, although the evening was chilly.

The following day, a fine one with billowy clouds hanging on the horizon, was spent getting our bearings. For lunch I wandered into a fast-food outlet called "Dixieland," where I was the only customer for some considerable time. However, after six weeks of eating a variety of Asian foods, many of them unknown to me, it was a pleasure to snack Western again for the first time. Back at the hotel I climbed to the hotel's rooftop to admire the view. Seoul is certainly a modern city—clean, full of flowers, with civic pride everywhere in the little details that make a tidy city a friendly one. It has been largely rebuilt since the Korean War of

1950–1953, when it was overrun by the North on at least two occasions and was severely damaged on many others.

In the afternoon we met up with Malcolm MacNamara from the New Zealand consulate, and he later took Russ and me to Nam Ch'un Chon to pick up John Edgar and Murray Brown. While there we also met Dr. Colin De'ath, an English professor at the Kangweon National University, in whose house we would be staying, and Mr. Kim Ju Han, who was the manager of the nephrite mine we were to visit. Together again as a team, we went out that evening to a restaurant, where there were 10 of us, including three Koreans. Food was cooked in a brazier in the middle of the table, and many toasts were proposed, as seems to be the custom in the East. Later we bedded down at Colin's house, and, as there weren't sufficient beds, I slept on the floor.

The next day we visited the local university museum, where there was a jade display. It should be apparent that by this stage of our journey I felt that I had covered more museums and seen more jade displays than most people will see in their entire lifetime. However, it must be said that visiting a museum is a valuable and efficient way of getting up to speed in a niche market such as jade, and by and large the displays and exhibits we saw throughout the East were at worst informative and, at best, very creatively presented.

Later Mr. Kim and his party arrived, and we set off, driving along clean paved streets and treed roads, for the jade mine, which was about 100 kilometres northeast of the city. On either side the rice and corn fields were ready for harvest, and the road that followed the North Han River was almost continually bordered by strips of flowers, making a very attractive scene. In the background the hills were green with trees.

From Chun-Chon we turned toward the village of Dong Myeon, which was about five kilometres south of the Soyang Dam. The reservoir, incidentally, was at one time the largest man-made lake in Asia and is a popular tourist destination. The mine turned out to be right next to a minor road through a narrow valley. Indeed, the tailings dump almost touched the verge. There were samples there for the picking, and Mr. Kim invited us to help ourselves, which we did, finding some nice pocket-sized specimens that were mostly light green. Also in the dump were crystalline limestone and gneiss and some altered lime silicates. On the north side of the valley were two other abandoned mines that had been started in an attempt to follow the jade structure across the valley from the south side, where our currently inactive mine was located.

We were then invited to go underground. The conditions there were such that the equivalent mine would be condemned in Canada, but in their defence it was not operational when we visited. Mr. Kim explained that it had originally been a talc mine, but in the 1950s someone from the U.S. Army had recognized the nephrite lenses, and the operation had shifted gears, so to speak. The mine's production had peaked about a decade before our visit and had closed some years previously, only to reopen when the price of jade rose sufficiently. In 1978 a subterranean stream had been crossed, necessitating pumps. These days the mine was left flooded, but it had been specially pumped out for our benefit. I picked up a nice piece of jade at the front face, but noticed that much of the material had stress marks, no doubt caused by blasting. In the long run, using explosives shortens a mine's life unnecessarily.

Afterward we visited a factory where necklaces and beads were manufactured from Mr. Kim's jade. We then drove to a restaurant, where a brazier was placed in the centre of the table again (in a hole, I should point out), and the food served hot from there. Like most Westerners, I found sitting cross-legged most uncomfortable and was pleased when we could get up and stretch. Mr. Kim then took us to his home, explaining while en route that the company had exported jade to Hong Kong, Taiwan, and recently five tons to Beijing, although it was not clear how they could still be shipping jade if the mine had been inoperative for so long. Later Malcolm MacNamara drove us back to Seoul. I noticed that at both the military checkpoints we encountered, our embassy car was given recognition and waved through promptly.

It being a Sunday, we declared October 5 a quiet day. My back was bothering me, probably from sleeping on the floor or sitting in an uncomfortable position in the restaurant the day before. As a result I chose to stay in that evening while the others went to an embassy dinner. However, I had the last laugh when the Kiwis returned to the hotel after the gate had been locked and were forced to scale the walls to get back in—shades of university days.

The following morning we visited the National Museum, which had been the Japanese administration centre during the occupation. The director, Mr. Gon-gil Ji, met us and provided some information on the jade collection. Later we viewed the exhibits and despite it being expressly forbidden to take pictures of the displays, we did anyway. I noted that there was some material on loan from the Royal Ontario Museum. Back at the hotel we packed up and left for the airport, where a plethora of armed guards, festooned with automatic weapons and other munitions, seemed to be everywhere. However, we passed through the system fairly smoothly, and in due course our plane lifted off for Hong Kong and the end of our three-country jade tour.

1994: Russia — Banished to Siberia

Almost immediately after the dramatic change in government in 1991 and the breakup of the USSR, it became apparent that it might be possible to get permission to visit Russia. Not just any part of Russia, of course, but Siberia, where some of the world's most historic jade fields lay. Once again my travelling associates were the New Zealanders Russell Beck and Dr. Alf Poole, with whom I had seen so much and travelled so far. Our fourth this time was to be Neil Lewis, another Kiwi, who was an avid jade carver. Since they were all from Down Under, I supposed that meant that I, as the sole representative of the northern hemisphere, was from Up Over.

Siberia's jade is the nephrite variety—the same as is found in New Zealand, Canada, Australia, and the United States. As in British Columbia, it was formed in association with serpentinites, although this was not a universal mode, with some originating from the metamorphism of dolomitic marbles (a site on the Vitim River, northeast of Lake Baikal being the best known).

As before, the trip was spearheaded by Russell Beck. It was his expectation that a new book on jade from around the Pacific Rim could be published, with major contributions from Russian geologists. With this in mind, an official invitation to visit Siberia was obtained, and we learned we were to be hosted by the Institute of Earth's Crust, Siberian Branch of the Russian Academy of Sciences in Irkutsk from August 4 to 19. So in due course I set out from Vancouver to join the group in Tokyo.

It was August 3, and I was lost in Narita Airport, trying to find three New Zealanders who were somewhere among the thousands of other travellers. Eventually we met up, more through luck than planning, and once together, we caught a train across the country to Niigata, whence we flew Aeroflot to Irkutsk in Siberia.

The plan was first to meet up with a group of Russian geologists, who were to be our guides. At the Irkutsk airport we were introduced to Eugene Sklyarov, Yura Menshagin, Dr. Alexander (Sascha) Sekerin, and Mark Sekerin (Sascha's son). For all of us, both hosts and visitors, I think it was a time of great expectations. Not simply were we going where few Westerners had ever been permitted to go, but the recent end of the Cold War had produced, suddenly and quite unexpectedly, a huge sense of relief and interest on both sides about the other society. The Iron Curtain that had kept us apart for half a century was gone.

Irkutsk (or UPKYTCK, as they would spell it), with a population of 600,000, is the second largest town in Siberia after Novosibirsk and is the chief administrative and economic centre of Siberia. A small, isolated town for centuries, its fortunes changed when the Trans-Siberian Railway (1891 to 1916) linked it to the outside world, and thereafter it grew rapidly. Our lodgings turned out to be student quarters and were third class, while the plumbing was fourth (out of order). Dinner was provided but wasn't very appetizing, and everywhere we looked things were very rundown and dilapidated. Clearly this was not an encouraging start, and Russell, Alf, and I steeled ourselves for a repeat of our Khotan experience.

With our hosts we discussed the itinerary until midnight before turning in. Then we listened to dogs barking until dawn (again, quite reminiscent of those times seven years earlier in Khotan), and spent the rest of a long night swatting mosquitoes. In the light of morning we visited the Institute of Earth Science, where the Director was Nikolai A. Logatchev. There was a collection of local jade on view, with both light and dark nephrite from dolomite and serpentine sources, which proved to be interesting. Later we were shown an experimental setup to change the colour of nephrite. The program was under the supervision of Dr. Vladimir Y. Medvedev. The details were somewhat secret, but I gathered it involved heat and pressure (200°C to 500°C and 500 bars). Apparently the colour was affected by the number of hydrogen ions present in the pressure vessel.

Later we had tea in a female geologist's office. Nina Sekerina was the wife of Sascha and a geologist too, specializing in nephrite. She told us that Siberia had an estimated 3,000 tons of nephrite reserves, although only two deposits were currently being worked. The fact that she spoke both German and English made communication a lot easier, and as always, I felt saddened that my North American upbringing had limited me to a single language, so that I had to rely on others to converse with me, rather than the other way around.

After lunch the work started. We began to load the expedition supplies into a large army-style truck, with double dual wheels on the rear. Pretty much everything that moved off-road in Siberia looks like it belonged to the army at some stage. It turned out that we had a lot of equipment. At the back of the building we hauled crates, and I couldn't help but notice six cases of Skol beer (24 to a case), lots of juice, and even more vodka. The Russian members of the group seemed concerned that we might not have enough "special equipment." It transpired that Olga (Olla), who was Eugene Sklyarov's wife, was coming along as the cook.

Somehow everything and everyone was finally all crammed in and, shortly before four o'clock, we hit the road, heading southwest and stopping at a village to buy bread. Along the way we passed fields of sunflowers, treed lots, occasional villages, and people appearing out of nowhere to wait at bus shelters or to pick berries at the roadside.

Near dusk we stopped at a viewpoint at the south end of Lake Baikal, which is both the world's deepest and oldest lake and contains an astonishing 20 percent of all surface fresh water on the planet. Considering how much of northern Canada is covered in lakes, streams, and rivers, that is a *lot* of water.

Lake Baikal's great depth (over a mile at its deepest) also means there are some very curious creatures, cut off from the deep ocean, which have evolved there since the lake formed 25 to 30 million years ago. Over 600 kilometres long and 80 kilometres wide, its average depth is over 600 metres.

Baikal is surrounded by predominantly rocky country, and despite having an enormous watershed (half a million square kilometres), there is little mineral uptake in the many rivers flowing into the lake, and at one time the water was renowned for its purity. Alas, that was before the arrival of Soviet heavy industry—notably pulp mills in the north. Only the Angara River flows out of the lake.

The cabins that were our destination for the night were located on a high point, and it took awhile to get there, as the trucks were slow. It transpired that some of Eugene's colleagues from the institute were in residence, so there was the usual backslapping and cheerful chaos. It was late and getting dark, so I decided to forego supper and get some much-needed sleep, as jet lag was catching up with me, and it had been a long day.

The morning brought drizzle. Breakfast comprised tea, bread, cheese, and strong sausage. We were adjusting culturally. The first plan of the day was to see a lazulite deposit—not to be confused with lazurite, also a blue mineral that is the principal colouring in *lapis lazuli*, a popular gemstone. Lazulite is a magnesium-iron-aluminum phosphate found in high-temperature metamorphic rocks.

Historically, boulders in the region had been identified in the river below the source and later traced uphill to outcrops, which had been exposed by a landslide in 1860. Back then, the area had been utterly remote, and it had taken two days on horseback to get to the nearest road. Today there are several small adits (horizontal shafts) and an open pit on the property, which is easily reached by road.

Calculated reserves were about 3,000 tons, but since mining was dictated by market demand, it was uncertain how long the source would last. Most of the production was exported, we were told. The material out of northern Afghanistan is said to be better. At the Siberian site some of the strata lodes were up to a metre thick, and the best grade was worth about $200 per kilogram. Pink sodalite was also found there. The resident expert was Dr. Eugene T. Vogovev, who explained the geology via Eugene Sklyarov's translation. I collected some nice specimens (nothing truly spectacular), but stopped when the weight became too much.

In the afternoon we visited a Buddhist temple and, later, another sacred site (I'm not sure which denomination), where the custom was to leave a gift for the god(s)—either cigarettes or vodka. I think the monks were on to something there. Then we pressed on southwest. In the evening we passed a village on the Irkut River and set up our tents amid mosquitoes and pines. Olla got busy with the supper, but for reasons that are now lost, I did not make a note in my diary of what supper was, which suggests that it may have been similar to breakfast.

The next morning brought a repeat of the day before. It was misty, and no one seemed inclined to get up early apart from the driver, who took the truck away for repairs. It appeared that we might be there for the rest of the day. Our breakfast was spaghetti, ketchup, ham, and herb tea. Nobody could accuse of us of not being omnivores. By midmorning the mist had slowly dissipated, revealing mountains to the north. The camp was at about a thousand metres, which explained the cool night, and the peaks rose

Organizer of the Siberian adventure, Russell Beck was curator of the Ivercargill Museum in South Island, New Zealand. His interest in nephrite was sparked by the strong Maori tradition for jade carving. It was his hope that the Russian visit would result in a new jade book being published, with contributions from Irkutsk scientists.

around us to 2,700 metres. Some of us started a short walk to view Silver Peak in the distance, which was reported to be over 3,400 metres.

The truck returned, and after lunch we set off again, passing through a village where there were numerous enclosures. It turned out these were fox fur farms (say that quickly after three vodkas). The sun was hot, and we stopped at another sacred site, where more vodka was consumed—by us, not the monks. Thereafter the road climbed north steadily, and along the route alpine meadows stretched, sprinkled with occasional grazing cattle. The pass's summit was at 2,000 metres, where the road forked, with one track leading to a gold mine. Time did not appear to be an important variable here, as we stopped to drink vodka with a grader operator whom we met on the road. Later at a lake a number of tracked vehicles were parked, and we stopped to have a few more shots with the occupants. No one seemed to be in a hurry to get anywhere fast. The afternoon slowly drew to a close, and at 7:30 p.m. we finally left. Sometime later we forded the Kitoi River, where we pitched our tents after failing to get into some cabins that were locked.

On August 7 it was sharply cold, and I felt shivery in the morning, and further, at an altitude of more than 2,000 metres, I found that I quickly got out of breath. As I was in my late 60s, I thought it prudent to avoid overexertion and to consume plenty of liquids, water included. By midmorning the sun had barely penetrated the valley mist, and we huddled around a fire.

Once again there did not appear to be any big hurry to get going. One of the locals showed off his rifle, which was some sort of AK-47 look-alike. By late morning the mist had largely dissipated, and there were no clouds above. Next to the camp, the Kitoi River was just a small stream, shallow and spread out over various channels. Our hosts were talking and slowly packing up, but I noticed the beer was out, and Olla was doing all the work. At noon, we were still waiting for the truck to come back with oil. However, the sun was warm, and we could strip off some of the outer layers. A pack train of horses arrived, and the outfitters conferred with our guides while more vodka was passed around.

Finally, at 2:30 p.m., our truck returned. However, another hour passed in linguistic confusion until we started off in a pair of tracked vehicles that looked like snowcats. In these we proceeded to cross and then recross the Kitoi River, always heading upstream in what was now an easterly direction. The machines roared, shuddered, slid and shook like broncos at a rodeo, and we clung on for dear life, envying those lucky cowboys who only had to endure eight seconds of this sort of motion. At about six o'clock in the evening, my teeth rattling and my muscles aching from bracing myself to cushion the shocks, our machine bogged down in a bank. Fortunately the second vehicle behind us could pass, and after much chaining and mud flying, it pulled us out. At 7:30 p.m. we stopped at some cabins that had the luxury of bunks. Apparently the buildings were once a hunting camp. I slept well, and there were no mosquitoes, which was a pleasure.

The morning was overcast and somewhat warmer than the past few days had been, which might have accounted for my good night's sleep. Breakfast started with an omelette followed by Russian porridge made from some unidentifiable grain and, finally, freshly caught fish. What an eclectic diet we had! There was no progress being made toward moving, and Sascha was not well. Alf Poole, as our resident medic, gave him some kind of pills. Meanwhile, we waited. One of our hosts explained that the local farmers found jade boulders in the river there, as we were not too far downstream of the source, and the bigger boulders they cut with a wire saw.

Finally, at 10:30 a.m., we packed up and boarded the tracked machines again, turning north up a tributary, and shortly afterward it started to rain. The howl of the windshield-wiper motor added to the cacophony of mechanical sounds coming from the tracks. Then the belt on the alternator broke and had to be replaced. When this was finally accomplished, we crisscrossed the tributary, which was full of boulders, and in between having our bones shaken by the constant lurching of the vehicle, I noticed that wild rhubarb grew in profusion along the track. Away from the stream the trees were tall but thin, and moss covered the ground.

The vehicle broke down again. The rain more or less stopped as the repairs went on, and shortly before noon the driver tried the motor, which turned over but wouldn't fire. It seemed there were more problems, so while the others were occupied, we "tourists" spent two hours hunting rocks along the Sagan-cir. We found some listwanite and serpentine but no jade. Back at the vehicles we ate lunch as the sun reappeared, after which we partly loaded the remaining working vehicle and went on to the jade camp. I remember that my boots were wet from the stream hunt, and I had no spares. However, Russell kindly lent me a pair of dry socks.

The litany of mechanical failures continued. Near five o'clock we were obliged to repair a track during a heavy downpour; undaunted, we got going again, the track held together with wire and a lot of faith. We limped along at a snail's pace, and it soon became apparent that we were not going to reach the mine by dark. That meant another night of camping out, as there were still eight kilometres to go. In the twilight we unpacked the vehicles. Dirty, wet, and tired, we rolled into damp beds in soggy tents.

In the morning it was sunny, thank goodness, and we packed up quickly, but the vehicle refused to start. This was becoming a habit. Neil suggested checking the spark plugs, which was done, and bingo, we were able to move on. Four hours later, after several halts and the addition of oil twice, we arrived at a gravel flat, where jade blocks from the mine were stockpiled. We were nearly there at last. I took some pictures, including one of so-called cat's-eye jade.

There was an old-timer in residence named Valery Kudgartseev, who had been the main producer of jade for decades. Sascha told me that he had first met Valery 30 years ago. We walked into camp to have lunch and later climbed up to a nephrite "vein," as my companions referred to it—I prefer the term "lode"—at 2,500 metres (8,200 feet). Graphite darkened the nephrite and the albite alteration. Sascha said that diamonds had been found in the serpentine, associated with the graphite.

As on the previous day, rain continued most of the afternoon, sometimes heavily. At six o'clock we loaded ourselves back onto the crawler and headed down the river valley. An hour later the downpour eased off, but there was more mechanical trouble, and it looked like we'd be camping again. Legs protruded from below the vehicle, while other bodies hung in and out of the engine area. Not understanding the language, we had to rely on the tones of voice to determine what was going on. There seemed to be a lot of cursing, which did not bode well, but then, unexpectedly, things were magically fixed, and we made it back to the cabins after all.

Those shacks might have been primitive, but they beat a tent any rainy night of the month. There was also an electric light (only 15 watts, mind you), and a sauna, so we could get clean. Dinner was a salad and caviar, a palatable stew washed down with the inevitable vodka, and wine spumante. I'm not sure what all the fuss about caviar is for—it's only fish eggs.

The following morning we left to visit another nephrite deposit while the second tracked vehicle was unloaded to facilitate repairs. After leaving camp and travelling for half an hour, we were forced to stop and clear a ford of tree trunks before we could proceed. After 15 minutes of log removal we re-inspected the ford and decided to walk the balance of the distance, because the crossing would likely be dangerous. Half an hour's tramp brought us to the site where there were some drill cores, and we were invited to take samples. Nearby there was a talus block of nephrite weighing some 50 tons or more. We later climbed a sidehill to a jade lode.

I made a quick geological sketch of the layout, and we all collected hand samples. Following a lunch of cheese, bread, sausage, biscuits, and tea, we set off for another quarry site down the valley, at an elevation of 2,100 metres. While there we inspected a nephrite vein up to two metres wide that had been drill tested to 30 metres. Sascha Sekerin said he had visited the site 20 years previously. I noticed that the rock was quite fractured, possibly from blasting but maybe from natural causes. Frost-heaved blocks were abundant downslope. Later we returned on foot to the ford, boarded the tracked vehicle and lurched back to camp.

August 11 started with another misty, damp morning. We were supposed to leave for Irkutsk the next day, so were going to hunt in the river and at the No.1 vein. But first we decided to wait for the weather to clear. At 11:30 a.m. the fog finally burned off from the valley, and we started out, but to my surprise I found I couldn't climb the first hill. We were at 2,000 metres or more, and I was "bagged," despite being just 69 years old. I decided I'd better not force anything, so I returned to the cabin to await the others' return. At two o'clock, Sascha's son Mark came in and got a fire going. On the group's arrival back, Valery suggested that Alf and I split the slab he'd taken from a small boulder, but Alf generously said I should have it all. Alf gave Valery a piece of his work, and Valery came back with another big block, which we arranged to cut later.

In the evening it began to rain again, with low cloud or fog. I did not feel well—shivery—whether from the damp or the altitude I was unsure. And to add to our discomfort, the following day dawned dull and rainy again. We left the jade camp for the last time, and after many stops for oil, track repairs, and belt replacements, we arrived at Fish Camp, where three of us were billeted in a cabin.

Wonder of wonders: on the following morning there was blue sky above the low-lying morning fog, and I had slept warm for the first time in a while. Breakfast was fish soup, with the saddest eyes I'd ever seen looking up at me from the bowl. Tough eating, but it takes a lot to put a geologist off his food. Sascha was ill however, and Alf Poole was consulted. It appeared that Sascha was unfit to travel, as he had food poisoning. In my diary I noted that I hoped it wasn't the fish. We declared it a rest day and stayed put, keeping Sascha warm in the cabin.

The next morning our patient said he was somewhat improved, so we loaded up the vehicles, but no sooner were we on our way than there was a disaster. Our machine became stuck midstream in the Kitoi River— balanced on a rock, with the tracks spinning ineffectively. Eugene Sklyarov waded ashore and found a tree to tie across the tracks, and after much mud wallowing and swearing we lurched forward and over the obstacle. The rest of the morning was spent in a series of starts and stops until we arrived at a lunch place, where a bull challenged our territorial rights. We ignored him.

Of considerable concern was the vehicle, which was not operating properly. Okay, when did it ever operate properly? At seven o'clock that evening we were finally rescued by a truck that brought in a couple of mechanics. We transferred from the tracked vehicle to the truck and followed the former to Cow Camp. On one further occasion the truck had to be pulled out of the mud. In this manner we continued until Monday Village, where we arrived just before midnight. We were taken to an empty house, where we gratefully slept on the floor, as it had been a long and bone-jarring day.

In the morning we took stock of our surroundings. The locals were of Mongolian heritage, and children on bikes gathered around, curious to see us. Neil Lewis tried to borrow a bike, but the boys were suspicious, so we walked around and took photographs to pass the time. It was certainly a charming village with interesting architecture. At noon we left, retracing our inbound route, stopping at a Buddhist shrine for some photos and then on again. Haying was in progress in the meadows beside the road, and it all looked very Third World, with scythes and stalk bundles, horse-drawn wagons and not a tractor or combine in sight. However, the roads were a great improvement on the previous week's tracks, and we made rapid progress. In the evening we stopped to buy beer at a Trans-Siberian Railway station and, a while later, paused at yet another viewpoint at the south end of Lake Baikal. Well after dark we finally reached Irkutsk and stayed over at Sascha's house. Alf Poole and I slept high (I had a bed, and Alf had a sofa). Neil and Russell slept low (on the floor).

In the morning we found there was no hot water and no mirror in the bathroom, which was something of a disappointment after the previous fortnight's facilities, and the toilet was in a separate stall. In short, the building was pretty rundown, but compared to where we'd been staying recently, it seemed very grand. Mark Sekerin took us to a nearby bank where, after much passport-waving and paper-signing, we eventually managed to cash some U.S. dollars. Then we sallied out for a smart lunch at the restaurant of the Intourist Hotel, where we discovered the walls were made of patterned gemstones that included rhodonite and lazulite. In the afternoon we visited the Museum of Limnology (freshwater lakes), and then, after stopping to see a local church, we returned to Sascha's house.

Midmorning on August 17 saw us at the Anthropology Museum, where we met the female director. Eugene Sklyarov had previously set up the appointment and made the introductions. It transpired that this was a 250-year-old building, which in itself was unusual, given the remoteness of the place in 1750. In 1869 a fire had destroyed over half of Irkutsk and damaged the museum too. The collections included Bronze Age artifacts from 6000 BC as well as jade adzes and scrapers from the Angara River near Irkutsk. I also saw 11 specimens from the Lena River, but they were of poor jade, mostly schistose. We were told that in the nineteenth century a local priest had collected Neolithic pieces throughout Central Siberia, and his collection had found its way into the museum. That is not to say that all the Neolithic artifacts had been found in remote locations; we were informed at the time of our visit there was a dig operating in the centre of a local Irkutsk park that was turning up jade artifacts.

Later there was a lengthy workshop by the New Zealanders on Maori society and its relation to nephrite. Then the Russians gave a similar presentation on how the aboriginal societies of Siberia used the stone. Both parties were very interested in the parallels of Siberian and Maori cultures. As the rain returned, we headed back to the Intourist Hotel. There we enjoyed a good meal—you will notice in this account how rarely this comment appears—and then we left for a jade factory in the village of Smolenstchina that also worked with lazurite and other semi-precious stones. Upon our arrival Sascha went to see if we were welcome, and apparently we were. In one of those surreal incidents that can only happen in strange places, the music of Scott Joplin was playing on a radio in the background as I chose a jade plate, an egg cup, and two small goblets, all quite reasonably priced, to take home with me.

On our way back we stopped off at the Anthropology Museum again, and then visited an old church where a flame burned continuously in the tower in memory of prisoners taken during the Second World War. Finally, we went back to Sascha's house, where quite a few of his family had gathered. Mark Sekerin played some Beethoven pieces on the piano after dinner.

It was sunny and warm the next day, and we went to meet the director of the Institute of Earth's Crust, Dr. Nikolai A. Logatchev. It transpired that he spoke English quite well, and we found him easy to talk to, although he was not overly interested in what we were doing there. Later, in Eugene's office, one of the technologists was given some of our rocks to cut so we could swap slices. Then back to the jade factory, where Neil picked up some jade mugs. Next we were driven to the private home of

Berber Woman *by Deborah Wilson, 7" x 4" x 2.5", jade*

Wilson's green and inscrutable Berber Woman *calmly fixes her contemplative gaze on something in the distance.*

Octopus *by Deborah Wilson, 6" x 10" x 21", jade*

Keep an eye out for this octopus. It was stolen from a downtown Vancouver art gallery in 2001, and has not surfaced since. It was priced at $25,000.

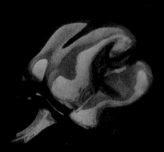

Green Pepper *by Deborah Wilson, 4" x 3.5", jade*

Looking good (and real) enough to eat, this pepper is one of several that Wilson has sculpted.

Buddha *by Lyle Sopel, 30" x 24" x 60", jade and bronze lotus base*

Starting with a two-ton boulder of grade A nephrite jade, Sopel worked over two years to complete this sculpture, the largest jade Buddha to be created in North America.

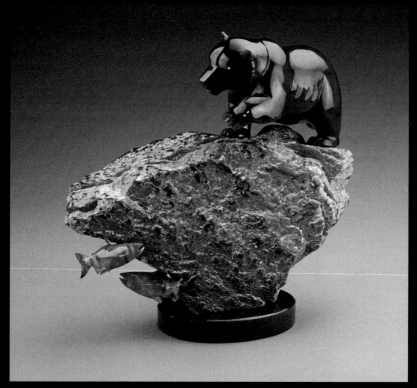

Two for Lunch, *grizzly and salmon by Lyle Sopel, 11" x 12" x 6", rhodonite and jade*

"I like the idea that the fish are hiding from the bear. To me, this captivating play between the salmon and the bear helps give this sculpture life."—Lyle Sopel

Journey Home, *Canada geese by Lyle Sopel, 22" x 12" x 9", jade*

Three beauties take off for parts unknown.

Cow & Calf, killer whales by David Clancy, 18″ x 16″ x 10″, jade

My Salmon, *standing grizzly by David Clancy, 24″ x 12″ x 12″, jade*

Northern Cohos *by David Clancy, 18″ x 8″ x 5″, jade*

David Clancy transforms frolicking whales, growling bears, and leaping salmon into sleek works of fine art.

Maya *by Alexander Schick, 7" x 6" x 4", jade*

The expression of Alexander Schick's charismatic "Maya" seems to change when it's viewed from different angles.

Erotic Leaf *by Alexander Schick, 1.5" x 4.5" x 8", jade*

After experimenting with wood carving, bronze casting, and clay, Alexander Schick became charmed by the hard smoothness of jade.

Hand *by Nancy Hadler Street, 14" x 6" x 5.5", jade*

Street used her own hand as a model for this intricate work.

Otis and Olive, the Flammulated Owls *by Nancy Hadler Street, each 8" x 4" x 6", jade*

Rare and indigenous to the west coast, flammulated owls are the only small owls with dark eyes.

Eagle *by David Wong, 3.5" x 5", jade*

Wong's carvings became distinctively Canadian very soon after he emigrated from China. At New World Jade Products in Vancouver, he trained new carvers.

Turtle *by Ruth McLeod, 5" x 4.5" x 3", jade*

This turtle was inspired by McLeod's childhood memory of how her little pet turtle moved and looked.

Northern Princess *by Tom Duquette, 5" x 5" x 4", jade*

This bust took Tom Duquette three years to complete. Duquette's work over the years has included acrylic paintings and sculptures in wood, ivory, bronze, marble and alabaster.

a retired geologist and mineral collector named Ravil S. Zamaletdiiniov. While we were there a Dr. H.A. Suturin arrived, with whom Russell had been dealing regarding the publication of Russell's new book on jade around the Pacific Rim. Dr. Suturin had an existing publication on nephrite (in Russian), which he gave to Russell. Our host, Ravil, gave us a piece of tumbled jade, said to be from West Sayan.

As the end of our stay was approaching, we started to settle up our financial affairs. I changed $1,800 (U.S.), for which I received 3,810,600 rubles. I was a millionaire, but alas, only in Russia and only momentarily. Since it was our last evening in Irkutsk, there was a party in our honour, and we had a great time swapping stories, proposing toasts, and drinking vodka.

On August 19 we were all packed up at Sascha's, but I couldn't find my six rolls of film. I'd searched my bags twice and combed the room too, but to no avail. It was time to leave. At the airport there was more tension and then a major panic when it turned out we were in the wrong section, but we finally found "foreign departures" and got through security.

We were flying on a Tu-154 to Khabarovsk. Lunch was encased in a plastic bag: bread, cheese, sausage, and a very sweet cookie. The coffee was a chicory-flavoured liquid in a plastic cup, and I noticed that some of my fellow travellers' cups had no handle—obviously salvaged from another flight. Alf came back from the toilet and warned against sitting down. We landed after three o'clock and walked to the arrivals hall, as there didn't appear to be any buses from the apron to the terminal building. From there it was a pleasant drive into the city, which then had a population of about 900,000, and curiously, there were lots of right-hand-drive vehicles on the roads. At the Intourist hotel there was no hot water, and the mattress felt like iron when I sat down on it. After dinner we had a walk by the river before turning in for the night, and Neil, who shared my room, claimed it was noisy during the night, but I never heard a thing.

At noon the next day we were back at the airport. Once through security, we had our first encounter since being in Russia with a clean washroom. It's funny how important the little things can become. Even the waiting room there was clean and bright. Later, on board another Tu-154, announcements in Russian and English stated it was 1,800 kilometres to Niigata in Japan, and once airborne, smokers lit up and stank out the aircraft.

1999: Australia — Black Jade Down Under

Although the aborigines have been in Australia for more than 40,000 years and have used stone for arrows and spearheads, awls, scrapers, and mortars, there is no archaeological evidence of their learning the ground-stone technique necessary to make use of jade. This was in spite of the presence of adequate supplies in two separate jade areas in the country.

The Tamworth Serpentine Belt was first recognized in 1962 by Colin Moore near Dungowan, some 40 kilometres east-southeast of Tamworth, in the northeast corner of New South Wales. Nephrite had formed there along the contact between serpentine and quartz phyllite of Silurian–Devonian Age. The contact was faulted, and the nephrite, commonly schistose, but good-grade material was reported to occur in lenses. Some 22 tons were shipped to Hong Kong buyers in 1971, closely paralleling the birth of the Canadian jade industry. Production from Tamworth has since been limited to a few tons a year for the subsequent decades, possibly overshadowed by the Cowell jade occurrences.

A report states that the colour varies from light green to bluish green to deep green, with low to intermediate translucency, and scattered grains of black chromite are present.[4] Chalmers compared the jade to New Zealand material and found it inferior in translucency and texture.[5] However, given the wide range of quality of the New Zealand material, it is difficult to speculate on what that means. Tamworth has been known for over 40 years yet has not become internationally recognized, so the indications are that the source is not of a particularly high quality.

Back in 1965 and halfway across the continent, nephrite was also discovered in the Cowell area of South Australia by Harry Schiller, a local farmer. Schiller collected a three-to-four-kilogram boulder of dense, hard rock near an outcrop of white dolomitic marble, and in early 1966 the specimen was positively identified as nephrite at Adelaide University. Cowell is located on the east side of the Eyre Peninsula, that large tongue of land that lies between Spencer Gulf on the east and the Great Australian Bight on the west.

In the decade following the discovery, the property has been the subject of disputes over ownership, and its product has seen a lack of market acceptance. Much of the jade is either very dark green or black. There is some more typical green material, but the developers attempted to champion the black. Certainly the material takes an excellent polish.

By 1986 all the deposits came under the control of Cowell Jade, which later went public as the Gemstone Corporation of Australia. The Department of Mines had an interest in the property and undertook a detailed study of the deposits that identified over a hundred outcrops. Some were quite small and were mined out over the ensuing decade, but others were very extensive.

The deposits are examples of the metamorphism of dolomitic marbles and calc-silicate rocks of Early Proterozoic Age, brought about by the intrusion of feldspar. The nephrite is fine-grained and in the form of nodular masses or fractured lenses. Reserves have been calculated, based on detailed mapping together with drilling to establish depth, and as a result, the Department of Mines estimates a body of 60,000 tons of recoverable jade—an impressive number. Canadians, by comparison, can only claim a fraction of that amount, although they routinely export 200 tons of material annually. It is therefore likely, based on current known deposits, that the Australians have one of the world's largest jade supplies. However, extracting the nephrite presents a problem, as the lode is contained in a hard host rock. This is in contrast to the easy mining in B.C., where nephrite is bedded in soft serpentine and can be removed using a bulldozer.

The Cowell material must be tackled with explosives, with the attendant possibilities of damage, but with expertise this can be minimized. Once a face is exposed, mining is done by the plug-and-feather method along a line of weakness, and the blocks extracted. This is, of course, a slow and expensive process. On the other hand, mining can take place year-round in Cowell, while extraction of nephrite boulders from Provencher Lake or Mount Ogden in British Columbia can only be done over a few short months of summer, when the snow recedes off the alpine slopes of the north country.

My timing to visit Cowell—I did not get the opportunity to see Tamworth in New South Wales—was most opportune, as two members of the Department of Mines in Adelaide were just about to visit the region. It was a day's drive to Cowell, rounding the top of Spencer Gulf. We checked into a hotel upon arrival and then went to visit the local nephrite-cutting operation, where one innovation was a specially built, Italian-made, wire saw. This had no doubt been developed for cutting very large blocks of marble but was equipped with diamond-charged ferrules strung on a wire to handle jade. For the much softer marble, cheap sand (silica), rather than diamond was used, enabling the wire to slice through the hardest marble in surprisingly quick order. At the mine we saw at least half of the more than about one hundred outcrops and gathered specimens while my department colleagues explained the genesis of the jade to me.

That evening we were invited to dinner at the home of Neil Smith, one of the original prospector–farmers who had acquired jade claims. Smith was also a mineral collector, and we exchanged some specimens, the most prized being a cluster of selenite crystals that he had found in the desert at a salt lake north of Cowell. It was so fragile that I ended up hand-carrying it on the bus and plane all the way home, where it now has a prominent position in my display case in Summerland.

The Australian jade story has its parallel in the Canadian story. Both have an excess of production and a limited market. In Canada, sales to Taiwan provided the initial solution. The material was acceptable and cheap—too cheap, some would say. On the other hand, climate has played an important part in Canadian jade mining, as the season is short (often less than four months), but the extraction process is easy, while in Australia the season lasts all year, but the host rock is hard and the process labour intensive. The Taiwanese demand provided the impetus for exploration in the Omineca and Dease Lake regions. With this expansion came the desire to develop a local carving industry.

By comparison the Australian producers found it difficult to gain acceptance of their dark material, as the colour was not appreciated in the Orient. As in Canada, they did try to foster a local market, and a few artists succeeded in producing some quality carvings. The common mistake was to assume that an expanding market would develop. Cash flow problems plagued most companies that got into the business, and as we are well aware, jade is not an essential commodity. If the price is too high, other "jades" are used in the jewellery and carving industries, and with every passing year, new and colourful semi-precious gemstones are being found and markets developed. Australia by itself has, in recent years, discovered and successfully marketed important new semi-precious deposits of mookite (a colourful silica); stromatolites (ancient fossils in jasper); tiger-iron (similar to tiger-eye from Africa); chrysoprase (a lustrous green quartz); and rhodonite (a pink manganese silica). All of them in their own way compete with the dark Australian nephrite for market share.

1999: New Zealand — Connecting with the Kiwis

My wife, Kay, and I had the opportunity to visit New Zealand in 1999, primarily to reconnect with Russell Beck and his wife, Ann, and with Neil and Cynthia Lewis, in Invercargill in the far south of South Island. It was on Russell's initiative that I had been given the chance to visit China and Russia and see those countries' jade deposits, and now I was visiting his home country for partly the same reason.

What do I remember about New Zealand? Primarily sheep, sheep, and more sheep, of course, scattered across green fields with few trees. I also remember the Kiwi accent, hot meat pies, and left-hand driving, which took some getting used to again (since during the war in Britain). And there were always a backdrop of snowy mountains breaking the western horizon.

One of the most famous localities for nephrite is on the west side of South Island, where it occurs sometimes as nodules and veins in serpentine and talcose rocks but is generally found as boulders. It was known to Maori as *pounamu*, or "greenstone," and was highly prized. They worked it with great labour into various objects, especially the club-like implement known as *mere* and the neck ornament, *hei-tiki*, that is widely worn by Maori on official occasions. By contrast the green jade-like stone known locally as *tangiwai* is bowenite, a translucent serpentine with inclusions of magnesite. The mode of occurrence of both nephrite and bowenite has been variously described in geology literature.

It appears that the Maori distinguished several varieties of jade, the differences in colour being caused by the variations in the proportion of ferrous silicate in the mineral. A pale-greyish variety was known as *inanga* and is the colour still favoured today. *Kahurangi* was a light green translucent form that was most prized, while *kawakawa* was jade of a darker green colour. According to Finlayson, the New Zealand nephrite results from the chemical alteration of serpentine, olivine, or pyroxene, whereby a fibrous amphibole is formed that metamorphoses through intense pressure and movement into dense nephrite.[6] New Zealand jade, called by early writers "green talc of the Maoris," was worked in Europe (notably in Idar–Oberstein, a famous gem-cutting region) as an ornamental stone, but recent land settlements have allotted all jade collecting to the Ngai Tahu tribe only, with the result that the industry is currently undergoing a quiet phase.

New Zealand, you will remember, is made up primarily of two islands, of which the southern one, cleverly called South Island, is the larger, but has considerably fewer people. Indeed, the farther south you travel, the lonelier the land becomes due to climate, the smaller the towns are, and the less seldom you meet anyone (other than sheep).

I was particularly keen to see a kiwi, that curious, flightless bird that is in many ways almost a mammal, although it does lay eggs—usually one or two massive ones, which the male incubates for 90 days. After hatching, the chicks emerge fully feathered and subsequently wander off with little parental input. It is small wonder that for the longest time these birds were regarded as a scientific hoax by Europeans. I discovered that the head of a kiwi is almost hairy rather than feathered, and the beak has nostrils right at the tip. The birds have almost no remnant wings, or even wing-stubs, but do boast an astonishingly well-developed sense of smell, which is handy for hunting their favourite food: earthworms.

Our destination of Invercargill lay near the southernmost point of the South Island, 46° south, and required considerable effort to reach. We flew into Christchurch, the largest city on South Island, on a rainy mid-October afternoon and the following morning, at 8:15 a.m., caught the "Southerner" bus hugging the island's east coast. The views of the sea were wonderful, the road good, the bus modern, and the style of travel a big step up from certain previous jade hunting trips to foreign parts. Shortly after lunch we passed through the university city of Dunedin, which bore a striking resemblance to many Scottish towns. It transpired that this was not surprising, given that many of that country's citizens arrived during Dunedin's gold rush in the 1860s. In fact, we learned that the name "Dunedin" is the Celtic word for Edinburgh, confirming the connection.

Our curiosity with names was further piqued when the highway swung inland sometime later, and we drove first through the town of Clinton, then changed buses at Gore. Just after five o'clock we reached Invercargill, a city of some 50,000, whose wide streets had been laid out in a grand style more than a century earlier. Being at the end of the line, so to speak, with nothing beyond except the relatively small Stewart Island and Antarctica 3,000 kilometres further south, Invercargill has been somewhat bypassed by the world's recent developments. Much of the charming architecture from its early boom times (when it was South Island's centre for meat and wool exports) is still intact.

The Becks met us at the depot and, after a day of being seated in a cramped bus, it was a pleasure to stretch, clean up, and just relax. Over dinner Russell announced that there was much to do during our limited time in the area, and the next day we would drive north to their cottage near the mountains.

The following day, shortly after lunch, we set out by car. Our route took us back to the town of Gore and then northeast to Tapanui, where the Blue Mountains provided a backdrop to meadows full of sheep. Everywhere gorse was in bright yellow bloom. Much of the country had been indigenous forest when European settlers arrived in the early 1800s, but large chunks of it were cleared to provide grazing. Now, slowly, reforestation was taking place on higher ground, such as on the Blue Mountains, where, alas, the clear cuts reminded me of my native British Columbia. Further north we joined the Clutha River valley, where apple trees were in blossom (October is springtime in the southern hemisphere), and hedgehogs and possums lay flattened on the road—an unfortunate sign of spring.

This was the region that saw considerable action during the gold rush of 1862, and many of the buildings from that time, such as stores and hotels, have seen little modification. In fact, there were small settlements that we drove through that, apart from the paved road, would have fit into a western movie set with their gable-fronted buildings, old signage, and corrugated iron roofs. Later we saw snow on the higher hills as we pulled up to the Beck's Arrowtown cottage under its canopy of plum blossoms. Shortly after arriving, Russell lit a fire in the living room to keep out the evening chill.

The next morning was sunny but cool, and we set off for the popular tourist destination of Queenstown, on Lake Wakatipu, where the higher mountain ranges start. Russell pointed out that on the other side of those peaks was one of the wettest places on earth, and he mentioned "tramping"—the curious word New Zealanders use for hiking. Given the rainfall in Fiordland National Park, "squelching" might have been more appropriate. The prevailing westerly winds, unimpeded by other land masses, circle the globe picking up moisture and then slam into New Zealand's Southern Alps, which have 18 peaks over 3,000 metres, dumping seven metres (1,800 inches) of rain on the outer coast annually.

To the west the entrance to Milford Sound is predominantly gneiss, with a narrow ultramafic strip where green bowenite is found. Although this mineral looks a lot like nephrite, it lacks nephrite's interlocking structure and resulting great toughness. It has a Mohs hardness of 6.0

and a specific gravity of 2.6. North of Milford Sound, the blue schists predominate, and there are several sources of nephrite, although collecting is no longer permitted. By comparison nephrite is quite different, having a hardness of 6.0 to 6.5 and a specific gravity of 2.9 to 3.1.

The head of Lake Wakatipu to the north is a well-known jade area. Another is on the west coast, at a similar latitude to Christchurch on the east coast, where the Taramakau and Arahura rivers flow down from the Southern Alps, cutting deep and impenetrable gorges to the sea. Both these rivers break through the chlorite schists that are heavily fractured and faulted within the mountain belts. It is there too, in a few isolated pockets, that nephrite has been found. Prior to a 1997 ruling that gave all jade collecting rights back to the Maori, boulders were hauled out of these gorges at considerable expense by helicopter. Today, there is almost no mining activity in the country, and a considerable amount of nephrite comes from elsewhere but is sold to locals and tourists alike.

When Captain James Cook visited New Zealand in 1775, he learned about the mythical greenstone that the local tribes described as being "a fish that was caught in lakes to the south." Since Cook made landfall on the northeast shore of North Island, it is likely that the locals he encountered had only the scantiest knowledge of the real sources of nephrite and were relying on legend. Possibly the original story of *pounamu* being found in Lake Wakatipu had changed with the retelling into the stone itself being a fish. This is now deeply seated in Maori mythology.

Our reason for visiting Queenstown was to reconnect with Dr. Alf Poole, with whom I had travelled to Russia and the Orient. Alf and Russell had a further connection by virtue of Alf being chairman of the Southland Museum board, of which Russ was the director until his retirement in 1999. Alf and his wife, Nan, welcomed us all into their home, which overlooked the lake. During lunch and much talk we watched the lake steamer carrying crowds up and down the reach. At the northern end of the lake the Dart Valley, which is in a special reserve in the Mount Aspiring National Park, is renowned for its nephrite. Collecting there is prohibited, but Alf kindly gave me a cobble of Westland jade as a souvenir to take home, as I had not seen much that I liked at Arrowtown.

Later we turned northeast through rolling hills and mountain passes to the town of Wanaka, located at the southern end of the lake of the same name. The community was surrounded by fine peaks, some of them still snow-capped after the winter. The whole area lies within the altered schists and amphiboles that occasionally produce nephrite in small lenses. The afternoon was warm and sunny, and at 5:00 p.m. we arrived at Brian and Rosemary Ahern's for another jade hunters' reunion. We sat in their garden, drank beer, snacked, and reminisced about our travels together to Asia in 1986. Later that evening we left for Invercargill again. Sadly, it was less than six months later that Brian died unexpectedly.

The following morning Russell dropped me downtown, and I walked home. There being no hills, the going was pleasant, although it was hot in the sun as midday approached. I passed the town's water tower, which is a local landmark, standing as it does above the suburbia surrounding it. A curious Victorian edifice of red brick and grey steel, it would not have looked out of place as the spire on an orthodox church on the Russian Steppes or as a lighthouse on some nineteenth-century British headland.

The next day we visited Russell's passion: the Southland Museum and Art Gallery, which is the major museum for the region and holds collections of international, national, and regional importance. Situated on the southern boundary of Queen's Park, the building took the form of a giant 26-metre-high white pyramid, styled on the Maori design for shelters in the Southland region. It featured eight galleries depicting the area's art and history—natural, Maori, and European. There was also a large live tuatara display—a reptile unique to New Zealand whose origin dates back to the dinosaurs. The museum had an active breeding program, and the animals were (and still are) a major tourist attraction.

We passed through the building's large foyer, which also housed the tourist information centre for Invercargill, and went through the art galleries. The Maori gallery displayed local pre-European culture and was, of course, of particular interest to me. On the west wall a spectacular example of worked jade was named "Kaoreore": a beautiful piece of rich green jade showing the sawing and polishing process with sandstone. Opposite was a 120-pound water-polished boulder on a plinth, the "Te Maori" touchstone. Display cases on the east wall contained personal adornment items such as *hei-tiki*, amulets, and pendants. Next to this was an exhibit of jade tools with a selection of adzes, chisels, knives, etc., and the process for working *pounamu* was illustrated. The weapons case on the south wall showed three *mere* (hand clubs), one of which had been featured on the cover of Russell's 1984 book, and there was also a very spectacular ceremonial adze in the case. Upstairs in the natural history gallery we came across the remains of a giant moa (now extinct) and a 300-pound water-polished jade boulder on a low plinth from the Livingstone Mountains in Southland. In the geology section were display cases of local minerals, fossils, and rocks.

As director, Russell was in charge of the entire operation, and his considerable expertise and interest in jade was certainly apparent. He had, over the previous decades, published a number of books and papers based on his research into the sources and uses of jade in early New Zealand cultures. One of his most interesting discoveries was the use of fire-treating jade. [7] In the late 1970s he had been puzzled by the unusual colour of some of the early artifacts in the collection. Instead of the usual green or grey, these were silver, brown, or even slightly orange. By heating rough pieces of nephrite in a fire, Russell discovered that he could recreate those colours, proving that some of the early items that had previously been thought to be "natural" must have been fire-treated. But why?

There was no doubt that some colours were more desirable than others in Maori culture. South Island was famous for its bright green translucent jade, but some societies prized a silvery-green variety known as *inanga* even more and went to a great deal of effort to find it, hidden as the source was in a very remote locality. It seems likely, therefore, that fire-treating was in part done for aesthetic reasons to produce a jade closely resembling *inanga*, but Russell discovered another important benefit: the heat process also hardened the stone. This was particularly noticeable in semi-nephrites, which are, before treatment, considerably softer than nephrite itself.

Over a number of years and many experiments, Russell showed that jade and semi-jade, heated to about 650°C for a short period of time, increased in hardness (as tested using the more precise Vickers Test rather than the Mohs Scale) by up to 50 percent. A temperature of 650°C was well within the range of a fire. Experiments using a modern furnace to push the temperature to 1,000°C revealed that beyond 650°C the nephrite began to craze or develop micro-cracks internally, which weakened, rather than strengthened, the specimen. The advantages to early craftsmen who laboriously worked the stone over many months were in being able to select the more readily available semi-nephrite, work it while relatively soft and then harden it and, finally, make it resemble the most desirable hue of jade. (This last point should not be underestimated.)

In the afternoon Russell took us to Orepuki Beach near Riverton to the west, where we spent a few hours beachcombing. It was a pleasant though windy afternoon, and I couldn't help feeling, as I looked out to sea, that this was the edge of the world with not a lot of anything in any direction over the horizon. It must have taken a great deal of courage for those explorers of centuries past—Tasman, Cook, Vancouver—to cross vast tracts of unknown ocean with absolutely no "Plan B" should anything go wrong. And then there was the uncertainty when making landfall; Abel Tasman, for example, lost several men when his longboat was attacked by a fierce Maori group in 1642, when he attempted to land at the north end of the South Island.

On October 20, we drove in light rain out to the Beck's cottage near Bluff, which is the port 25 kilometres from Invercargill. Ann told us Bluff is noted for its oysters when in season. Their house was set in a large garden filled with interesting and unrecognizable (to me) indigenous plants, in which both Ann and Russ took a great interest. Close by, Omaui and Oreti beaches provided miles of sand on which to stroll and berms of interesting pebbles to sort through while looking for minerals. It was a pleasant end to our time in New Zealand, and I was pleased we had made the effort to get together with my former jade companions—the more so in light of Brian Ahern's untimely death shortly thereafter. While we had certainly covered a lot of miles just getting to and from the main thoroughfare of North Island, my particular thanks went to Russell and Ann for making us feel so at home and for arranging a wonderful trip around the region to see so much and to visit old friends.

Jade in Canada's Courts

Over the years there have been a few cases of the Queen versus John Doe (and sometimes the reverse), involving jade in British Columbia. I was involved in one case, but, I hasten to add, as an expert witness, not as the accused.

~ The Cassiar heist ~

The case in which I was called to act as a witness involved the theft of 30 tons of jade. It happened at the asbestos mine in Cassiar in 1980. This small village was built for the employees of the asbestos mine that produced prime fibre from a deposit on McDame Mountain. For many years the hard toughness of the jade rocks were a nuisance to the miners, who needed the long-fibre asbestos, so the hard rock was added to the waste dump. Just how much was there is unknown, but eventually it was discovered that this waste rock was more valuable than the primary product, as mentioned in a previous story. Because the mine was an open pit, the area that hosted the jade was blasted, and the jade removed about once every two to three years. In the spring of 1980 that area was being worked, and it was the company's standard practice to mark the jade boulders with red paint and send them down the mountain to a secure area at the base.

In late August of that year a conservation officer noticed a pile of jade boulders in a ditch alongside the highway, about 40 kilometres south of Cassiar. A couple of days later, he noted that they were still there and reported the matter to the local RCMP. On September 6, Corporal Van Acker arrived to check out the situation, only to find two men in the act of loading the jade into the back of a truck. Their explanation as to why they were doing so, and whose jade it was, proved evasive; Ray Empereale and Clark Monteith were subsequently charged with theft of a value exceeding $200 and possession of stolen property of a value exceeding $200.[1]

In due course the case went to trial, and I was called as an expert witness to testify for the prosecution. In May 1982, I flew north to Watson Lake and caught the bus to Cassiar. Snow was still on the ground, and most of the lakes we passed were frozen. I was billeted at the mine camp, met the prosecutor and then spent a restless night. This was my first exposure to the law in any capacity—as accused, witness, or casual observer—and I didn't want to be made to look foolish by the defence. In the morning a number of us arrived at the recreation centre, where the circuit court was in session, with Judge Campbell presiding. The first case involved some Natives from Good Hope, but by 9:15 a.m. no one had shown up, so our case started. The two accused had chosen to be tried by magistrate alone rather than by judge and jury, which I found interesting. I assumed they had chosen judge only because jurors from within the community would likely be mine company people apt to favour the company's interests.

As a witness I was not allowed to listen to the testimony given by others, but I later obtained a copy of Judge Campbell's ruling, and it makes interesting reading. It seems that both men were involved from time to time with the collecting of jade at the mine, and Ray Empereale had even been involved in selling jade on behalf of the mine in 1977 when the last "crop" of jade had been salvaged. Over the course of the trial, the prosecution produced a string of witnesses who said they had seen the two accused loading jade boulders into the blasting truck (Empereale was a blaster by trade), and one or two had even told them to stop. However, the defence produced evidence that both men were expected to handle jade as part of their respective jobs.

How the jade got from the mine, which was in a secure area, to a spot in the bush where they "found" it, was explained by the prosecution. There was an extremely rough back road that was accessible to anyone, known as the Old Cat Road, and down this track, it was contended, the accused skidded a "stone boat" (a steel sledge) using a tractor. They then dumped the sledge in the bush not far from Shell Oil's camp. From there they later collected the boulders in four-wheel drive trucks and stored them in a ditch next to the driveway of Ray Empereale's house.

The defendants argued they had "found" the jade while out horseback riding, that it was obvious it had been there for a period of time as it was overgrown and that they were entitled to it. It took them three weeks, working evenings and weekends, to manhandle and sling 30 tons of jade boulders into the back of their trucks. A Shell Oil employee confirmed that he had seen their vehicles passing the camp on several occasions.

The issue was, had they stolen the jade themselves, or did they know it was stolen? The law is a funny thing and grinds along at its own pace. They'd brought me in as a witness to establish whether the jade was or was not from the Cassiar pit, and I took the stand for about 45 minutes. Since Cassiar jade has a characteristic chrome garnet in it, I was able to say that there was a 60 percent probability that it was from the pit. The fact that there were no other sources for miles around, that the company was in "jade mode" at the time of the heist, and that the jade was blasted and not drilled all pointed to it being Cassiar jade. Finally, there were red markings on the boulders that were consistent with the red paint used to mark Cassiar jade at the pit.

Both the defendants tried hard to show that they were just regular guys, with no real experience of jade, who were unaware of the value of "that old pile of rocks" out behind the Shell camp. I was asked for my professional opinion as to what the value of the stolen pile was, and I offered $1 a pound, making it worth $60,000. The prosecutor restated the value, saying that at $1 a pound it would be worth $60,000, but at $2 a pound it would be worth $120,000. I commented that there was nothing wrong with the arithmetic, which caused some amusement in the court.

Judge Campbell noted in his summing-up that, as both the accused claimed they did not know the jade's worth, credibility was being stretched to the limit. Both men had worked for the mine for a number of years, had been involved in day-to-day jade collecting, and Ray Empereale had been involved in the sale of jade on a commission basis. There was no doubt in the judge's mind that the accused knew exactly what they had stored in the ditch under bushes outside Empereale's home near Pinetree Lake. As well, the two men had not checked whether the area was under claim, in which case, old or new, the "pile of rocks" belonged to someone under law.

The trial continued for a second day, but I was finished and allowed to leave. Large snowflakes were falling as the police cruiser gave me a lift to Watson Lake. North of Boyer Lake a bus was stopped next to a man lying at the side of the road. It turned out he wasn't dead, just sleeping. However, the Mountie had to return to Cassiar, so I caught the bus to Watson Lake and then the plane south.

A month later the judge brought down his ruling. Both of the accused were found guilty of possession of stolen property, but the charge of theft was dismissed, as the judge felt there was reasonable doubt as to how the jade actually ended up in the bush behind the Shell Oil camp.[2] The men got two years each, but Clark Monteith was not present at the sentencing and a warrant was issued for his arrest. In an out-of-court statement afterward, Judge Campbell said he felt that, considering the importance of jade in the Cassiar community and the hundreds of thousands of dollars involved, prison terms were warranted. In retrospect, having seen the immense amount of effort, energy, and taxpayers' money that went into the case, it might have been more practical to have recovered the jade, have the mining company fire the offenders from their jobs and let the dust settle. However, the arm of the law is "long," and the mill of the law grinds exceedingly fine.

~ Jade artworks stolen ~

Like any valuable, small object, jade carvings have from time to time been the subject of interest to thieves. In December 1972 a jade heist took place during which locks were forced on display cases at the Hotel Vancouver and $8,000 worth of jade vanished.[3] As well, in March 1975, criminals broke through a bathroom wall at the back of White Jade House in Vancouver and took $40,000 worth of carvings and other artifacts.[4]

And then, in 1983, there was the curious matter of the theft of a jade bust that had been on consignment for two years at the Gallery of the Arts on Howe Street in Vancouver. The selling price was $3,000. However, about a week before the item vanished, the artist had visited the store, explained that he was short of money and suggested they drop the asking price to $2,800. A week later the theft was reported to the police, who interviewed the artist at some length as "a person of interest," but no charges were laid, nor was the bust ever recovered. I am not sure whether the insurance was ever claimed.

~ Jade causes air crash ~

A sad account of how jade played a key factor in a plane crash was reported in the *Vancouver Sun* in September 1971.[5] A twin-engine Piper Comanche piloted by a Denver, Colorado man crashed shortly after takeoff at the Prince George airport, killing both the pilot and the passenger—the latter a member of the University of Colorado basketball team and, therefore, likely just out of his teens. According to an RCMP spokesman, the plane had on board 1,200 pounds of jade acquired in the Fort St. James area. The plane was only certified to carry a maximum load of 1,000 pounds. It appeared that the loosely packed jade may have shifted shortly after takeoff, flipped the aircraft over and plummeted it to the ground near the runway. This was the first air accident at the airport in 15 years.

~ The Stein affair ~

In the early 1970s a prospector, as I shall call him, had several mineral claims on Skihist Mountain, above the Stein Valley, not far from the Fraser Canyon and the town of Lytton. It was his stated opinion that the properties were home to a deposit of nephrite containing "some of the purest jade in the world." The site was known as Green Gold. Unfortunately for the title holder, the provincial government placed the area under a mineral moratorium while the valley was being considered as a park, thus preventing any mining activity. The owner attempted to sue the Crown because of the confiscation of what he claimed was very valuable property. In fact, he said he had turned down a $2 million offer for the claims and had negotiated a deal in which he would receive $500,000 a year for the next 45 years.

In a *Vancouver Sun* article the prospector stated he had an agreement to sell the mine for $23 million if an access road could be completed by January 1, 1981.[6] In due course there was an out-of-court settlement between his company and the province of B.C.

I wonder how journalists can publish statements such as those that appeared in the *Vancouver Sun* of April 14, 1978: the owner was quoted as saying that until 1974, he had removed 100,000 tons of ore from the claims. It takes some imagination to believe that so much material could be removed, especially without an access road, which was deemed essential to the sale of the property. As a further aside, a simple check at the local mineral titles office would have revealed that, although the prospector thought he had a deposit of jade, the Green Gold site was already known for the rock idocrase (at one time known as vesuvianite), a green mineral that sometimes resembles nephrite. Although it is an attractive lapidary material, it is neither jadeite nor jade.

~ The Mohawk theft ~

Mohawk Oil's mineral exploration arm at Kutcho Creek was normally guarded by watchmen in the off-season, but even they don't see everything. In November 1982, during transport of a 19.5-ton block of jade and a 30-inch diamond saw blade from the mining camp to the highway at Dease Lake, a theft occurred. Now, one jade boulder looks much like any other jade boulder, but diamond saws are marked with a serial number. That was the undoing of the culprits.[7]

It turned out that two of the thieves tried to cut out a third, who subsequently tipped off a Mohawk manager. The guilty parties were quietly followed, resulting in the recovery of the stolen items in a subdivision of Kelowna, where the 6 x 6 x 8-foot boulder was in the process of being cut up when the two men, both in their late 20s, were apprehended. Charges were laid.[8]

As an interesting legal aside to this story, it was decided that although the jade boulder was, in effect, "Exhibit A," it was not necessary to produce the actual boulder in the courtroom (what a relief). Instead, a photograph was produced, and witnesses under oath were asked to attest to it being the actual boulder.

~ Placer versus lode ~

Two distinct forms of mining are recognized in B.C. Placer mining is "the winnowing of desired minerals from unconsolidated deposits." That means you can sort loose sands and gravels (usually with a pan) to find what you are looking for—usually gold, other dense metals, garnets, and similar. The other form of mining is lode mining, which involves "drilling and blasting in the hard bedrock to liberate the desired mineral."

As a result of this separation of roles it is possible for two miners to occupy the same property, with accompanying opportunities for conflict. One such relationship was between two jade miners in the Dease

Lake area. It so happened that there were jade boulders in the creek, downslope from the lode. The placer miner claimed they were his; the lode miner said they were his. In a 1980 B.C. Supreme Court trial, Mr. Justice A.A. Mackoff ruled in favour of the placer miner, Arnold Racicot of Richmond, that he had the rights to the loose boulders and that the lode miner must stay clear.[9]

~ Geyer's theft ~

This was a strange case, as most of them are. As reported in the *Vancouver Sun*,[10] Gunther Geyer proposed the construction of a $3 million centre for the arts that would include a school for jade carving. Mr. Geyer was best known for his life-sized statue of a rearing horse, carved from a large block of jade and said to be eligible for entry in the *Guinness Book of Records*. "Dr." Geyer is quoted as saying his family would put up the initial money for the centre.

In due course the project came to nothing. But in 1977 an alleged robbery occurred, in which many of the carvings belonging to Mr. Geyer disappeared. There was doubt regarding the authenticity of the heist, and the insurance company subsequently refused to cover the loss. Gunther Geyer took the matter to court, and I was called upon as an expert witness. The trial was put off because of Mr. Geyer's ill health but finally set for April 1979, when Mr. Geyer made a motion, asking Mr. Justice Bouck to disqualify him from the case on grounds of bias. This the judge refused to do. At that point, Mr. Geyer and his family declined to proceed, and the action against the insurer was discontinued.

The Jade Sculptors

Sculpture is an art form that creates representations of real or imagined objects in three dimensions. There are two styles: in the round and in relief. Both can vary from extremely small to huge.

There are two methods of sculpting: one in which material is built up (the additive process), best exemplified by metal sculptures, and the other, the subtractive process, where material is removed by knife, chisel, or grinding wheel. With the additive process, there is almost no limit to the size of sculpture that can be created. However, with the subtractive method the limit is imposed by the size of the initial raw material that is available. Marble or granite can be quarried as large blocks, but jade is a different matter. First, it would be difficult to get a large piece of flawless jade. Second, the cost would be very high. Third, the challenge of handling a large block would be prohibitive. Finally, the ultimate cost of the piece would be exorbitant.

Carving is more or less synonymous with sculpting but seems to be a less imposing term. "Sculptor" sounds grander than "carver," but this may smack of snobbery. The two terms are used interchangeably here. Art, on the other hand, is difficult to define but seems to require two faculties: manual skill and insight. A skilful craftsman could reproduce a sculpture such as Michelangelo's *David*, but only a true artist could conceive and execute similar work.

Jade is considered one of the most difficult mediums to use. It cannot be shaped with a chisel, and saws have a very limited ability to perform the work. Excess material must be removed by the abrasive power of the grinding process. While jade is not as hard as some of the minerals found in granite, for example, its great toughness prevents flaking or chipping—a good thing when finishing a work but one that slows down the shaping process considerably.

Jade carving in British Columbia as an art form is very recent and can be traced back to a visit in 1972 by the sculptor Robert Dubé to the manufacturing plant called New World Jade in Vancouver. He had recently bought an Eskimo soapstone carving from Marion Scott, who had an art gallery on Howe Street in Vancouver. He took the piece round to show the principals at New World Jade and posed the query: "Why not do the same in jade?" Why not, indeed.

Dubé proceeded to demonstrate how it could be done using the few simple pieces of equipment the company had in the plant. He needed only a small diamond saw, two or three grinding wheels, and a polishing machine to illustrate the technique. Thus was born the idea for New World Jade Products, a subsidiary of New World Jade Limited.

Robert Dubé was born in St. Jean Port Joli in Quebec and received his art training at the Minneapolis Art Institute. He was to become one of Canada's best jade sculptors, but he learned his craft in the traditional Quebec style of intricate woodcarving. A wild character who created sculptures that were larger, or more detailed, than anything else at the time, he was best known for a life-sized bear carved from B.C. jade. Very intense, Robert would work himself to exhaustion and complete pieces in a surprisingly short time. His attention to detail and the subtlety of lines made his work, like that of Deborah Wilson's, stand out. He also created many stylized animal shapes, which were popular with collectors nationally and abroad. Robert subsequently moved back to Montreal, where he died in the late 1980s of pneumonia at the early age of 38.

At about the same time a talented Chinese immigrant, David Wong, was in Canada travelling across the country, sponsored by Birks Jewellers. Wong had a contract to give jade-carving demonstrations and promote jade awareness for the store. He had been born in Kwang-tung Province in China at a time of great strife: civil wars were followed by the Japanese invasion in 1937. In later life David remembered playing games with young Japanese soldiers. Both of his parents were artistic, as is so often the case with talented people, his father being a noted painter and photographer.

At the end of hostilities in 1945, David moved to Canton to join his brother, an ivory carver, who taught him the art of creating sculptures. David subsequently served an apprenticeship in Hong Kong for another five years, after which, with support from his church, he went to university in Taiwan to study law, planning to be a diplomat. In 1966 he qualified to sit for bar examinations, having graduated from the National Chengi Chi University. However, it quickly became obvious that law was too political for David. The old creative urges, plus his obvious business orientation, were to pull him in a different direction.

Jade was discovered in Taiwan in 1964, but there were few jade carvers in the country. Further, the quality of material was poor. That same year, David Wong watched two hired boys pull a wire and feed grit to split a 10-ton jade boulder that had come from Canada. Seeing that single boulder had a profound influence on David's life. He knew what he wanted. And that could only be found in Canada.

In 1971 he arrived in his new country full of hope but unemployed. At the Pacific National Exhibition Gem Show in Vancouver he met Mr. and Mrs. Fear of Tsawwassen, who were amateur lapidaries. They had a well-equipped basement, and David was invited to use their facilities, which he did for a few months. It was then that he signed up with Birks, and at the end of the contract David decided to apply for landed immigrant status, after which he brought his family to Canada. That same year he met Robert Dubé. They worked together for a while, and in 1972 New World Jade Mines, through the efforts of Mr. T. Robert, started a subsidiary called New World Jade Products to process some of the fine material New World Jade Mines was producing from Mount Ogden.

A prior attempt had been made in 1971 to start a local jade processing industry using Canadian material, when Kuan Yin began to manufacture jewellery in North Vancouver; the venture proved unsuccessful, and New World Jade Products took over the premises. In 1972 David Wong began carving for New World Jade Products, and shortly thereafter, the company moved to Gastown in the east end of Vancouver. While there David began training new carvers. Within a few months there were 25 employees: carvers, trainees, and technicians. Stan Schmidt helped in the development of tools and techniques. The company employed numerous graduates of the Vancouver School of Art (VSA), now the Emily Carr Institute of Art and Design. These included Maureen Morris, Alex Schick, Deborah Wilson, and others. One of the first pieces to be produced was sold by Marion Scott from her gallery.

There was no doubt that a high-end market for quality jade carvings was developing. With the growth in sales, more artists were employed. Robert Dubé left New World to set up another studio, at first in the vacated premises of Kuan-Yin Ltd. in North Vancouver, then later on Cordova Street in Vancouver's Gastown area. There the new company produced hundreds of jade carvings.

Initially training costs at New World Jade Products were underwritten by Canada Manpower, a federal government agency at that time. Mind you, not all trainees developed into consummate artists. It was a new era in a new art form, but numerous works from that period did end up being owned by prestigious institutions and collectors across the nation and, indeed, around the world. A number of sculptures were bought as gifts for visiting heads of state and for other important out-of-country guests. The studio produced a pinnacle of modern West Coast art, in much the same way as Haida Gwaii artist Bill Reid's sculptures symbolized another line of recognizable artistic excellence.

However, a year later high overhead and small production drove the company into receivership. Two principals in New World Jade Company had such faith in David Wong, though, that they offered to set him up in his own studio and take over the marketing of his output. By April 1975 this venture also foundered, and David began his own marketing, mainly in Victoria and Toronto. Times were tough. From 1975 to 1976 David attended numerous gift shows, both to display and to sell his art. He survived financially and later moved out to Surrey where, with new agent Norman Doyon, he prospered in a studio called David and Kwai Jade Arts. David died unexpectedly of a heart attack in January 1998.

New World Jade Products was not the only show in town. Arte de Jade was the brainchild of Howard Lo and Dr. Glen Kong. Like the others, it was a company geared to produce high-quality carvings, and Robert Dubé ended up working with them. It was there he carved his *Big Bear*, a life-sized sculpture of limited artistic merit but no doubt a major technical challenge to complete.

Elsewhere, Jade World harnessed the skills of a number of carvers from Asia. Arnold Chow was trained in plastics manufacturing in Hong Kong but readily took to jade carving when he immigrated to Canada in 1974. Jason Ho hailed from mainland China, where he trained in traditional jade arts before relocating to B.C. in 1980 and going to work for the Jade World studio. The company tackled some fairly unusual carving challenges. Abraham Su was a talented carver who once produced 200 individual killer whale sculptures as gifts for attendees at an economics conference in Davos, Switzerland. Elsewhere, a group of sculptors from Jade World combined to create a jade staircase for the Sheraton Hotel in Anchorage, Alaska. The final cost of that project is unknown, but it required an immense amount of creative and manufacturing effort.

Sadly, Jade World closed its doors in 1988, a victim, like so many other businesses, of low labour costs in China. Even with its demise, some of the artists were not ready to give up. They formed a co-operative that provided space and equipment for artists Deborah Wilson, Ruth McLeod, David Enn, David Reuben, Alex Schick, and David Clancy. And throughout all the formations and consolidations, Don and Gwen Lee of Lee's Jade & Opals in Langley provided the pick of each year's jade material from the north so that artists could work with the best quality available.

Alex Schick was a Vancouver-born lad who trained, like many of the others, at the VSA, where he went through the gamut of woodcarving, bronze casting, and clay, but he was charmed by the hard smoothness of jade. One of his pieces was bought as an official gift to the government of the Bahamas, while another is in the national Museum of Man in Hull, Quebec. He now operates from his home in Richmond in the Lower Mainland. While he still carves jade, he also does large sculptures in soapstone and marble.

David Clancy was born in Etobicoke, Ontario, but like the others found his way to Vancouver and the VSA. In 1973 he found himself working at New World Jade Products for the equivalent of three dollars an hour plus a 5 percent commission. From his point of view that situation couldn't last, and it didn't. A great artist, Clancy was not afraid to tackle large blocks of jade, including a 700-pound free-form block from Mount Ogden. His sculpture of a bull goring a bear (a symbolic image for those who follow the stock market) won him critical acclaim and a higher profile in the Canadian corporate world, where serious collectors existed. However, most of his later buyers have been American.

David Enn was born in London, England, and arrived in Canada as a landed immigrant in 1965. He too went through the VSA before wandering the world for a while, returning to B.C. to try his hand at soapstone carving. The medium suited his style, and he moved on to jade at New World, producing many animal forms under the name of David

Nesbitt, signing his work "DN." Like the others, he subsequently moved to his own studio, where he produced sculptures that have found themselves in international collections, as well as in one of my own.

In its turn the co-operative split up, and the final holdouts formed The Jade Gate, a small premise in Vancouver where Deborah Wilson and Stan Schmidt worked, with some input from Alex Schick, Lloyd Manuel, and Tom Duquette.

Lyle Sopel opened his own studio in Kerrisdale in Vancouver under the name Jade Expressions, after spending time at Arte de Jade. Later he moved to North Vancouver and, in an effort to increase output, employed assistants to do some of the precutting and polishing. Lyle continued to do the designing and over the years has undertaken some very large-scale carvings.

It seems hardly surprising that there were, and are, so few women carvers. Despite the elegance of the finished product, jade carving is often extremely dirty work during the shaping process and, more importantly, is physically demanding. As a result, it has not attracted many of the fairer sex.

One of the few women jade carvers was Nancy Hadler (later Street). Unlike many of the others in the New World group, she had no formal art training but learned as an apprentice to David Enn and quickly showed herself to be a masterful sculptor in the jade medium—developed, no doubt, from her early woodcarving skills developed as a teenager. After the studio dissolved she moved to 100 Mile House in the Cariboo for awhile but was last reported to be on Vancouver Island.

Another woman who passed through New World Jade Products was Ruth McLeod, who trained at a number of art colleges in Canada and the U.S., graduating from the Ontario College of Art some years after working with jade in 1972. Although she did not produce a great deal, her choices of subject were often memorable. Her birds were highly stylized, yet they captured the bird essence perfectly, and her leaf motifs were haunting in their realism, compelling the observer to touch and feel the cool curves and ripples of the pieces.

In terms of both quality of work and longevity in the field, Deborah Wilson must be recognized as one of Canada's leading jade carvers. Like many others, she graduated from the VSA and then worked at the Jade World studio. Numerous articles have been written about her and her output, and she has produced an astonishing body of work over the years. She is master of the curved form, her carving being at times anatomically accurate to the finest detail and at other times producing abstract art so smooth, so folded, so curved, as to demand to be touched. She now works from a studio in Vernon in the interior of B.C. Her work has been acquired by collectors on three continents, and she is actively producing commissions for public works, galleries, and collectors. She also lectures and demonstrates jade carving and, since 1998, has hosted sold-out sculpture clinics, where aficionados can come to learn the basics of carving in nephrite. She also does quality work in other semi-precious stone such as lapis, marble, and agate, mainly on commission.

One visitor to the Vancouver jade studios was David Piqtoukun, an Inuit from the Northwest Territories, who spent a year adapting his soapstone carving methods to that of jade. The hardness of jade, of course, necessitated the use of grinding wheels and abrasive powders, unlike in the carving of soapstone. Much of his time was spent getting a final finish, something that was not as time-consuming in soapstone work. He initially attempted wildlife figures but found the proportions difficult to master. Compared to the easy expressiveness of soapstone, which suited the rounded figures of Eskimo art, nephrite could be reduced to spindly shapes—legs, horns, detail—that allowed a whole new world of experimenting. However, after a year in the south David was anxious to return to his home, where he continues producing works in varied mediums that include important commissions of public art for Canadian cities.

Of the 30 or so original jade carvers, there are about six working in the medium today.

The largest sculpture made from B.C. jade is almost certainly the *Jade Buddha*, which comes with a number of curious stories attached to it. In Bangkok, Thailand, a dynamic monk by the name of Phra Viriyang had for over 30 years been building a monastery to house sacred relics of the Living Buddha. A master manager and arranger, Abbot Phra Viriyang raised funds to build not one but 13 temples and a hospital in that country and another five in different cities in Canada—all this from someone who had taken vows of personal poverty.

It must have been the Canadian connection that set his mind thinking along the lines of a great project. He visited Vancouver during Expo '86, where he learned about British Columbia's great jade production. In 1987 he is reported to have had a dream, or a vision, in which he saw a giant stone Buddha made from the most perfect material. It had to be jade. The good abbot subsequently visited B.C. several times and urged his contacts to keep an eye open for a jade boulder of suitable quality and size, but there were none.

Years passed, and then in November 1991 he had another vision, in which he saw a great jade boulder close to a river. Subsequent inquiries to Jade West in Vancouver revealed that a giant, 32-ton boulder had recently been unearthed just a few metres from a stream on their Kutcho Creek property.

Phra Viriyang flew to Canada, where he and Kirk Makepeace, chairman of Jade West, closed a deal worth $350,000 (U.S.), unquestionably the largest price ever paid for an item of nephrite. Then, while the rock was being shipped to Bangkok, the Abbot flew to Carrera in Italy, home to some of the world's great marble cutters (Michelangelo worked mainly in Carrera marble). While there he sought to contract master sculptors capable of releasing the sleeping Buddha from within the great green rock. Two men were duly appointed, but they had no experience with jade, only with the much softer marble. (Jade has a hardness of 7 on the Mohs' Hardness Scale, while marble is just 5, meaning jade is at least a hundred times harder.) Adapting to the new medium, they spent 18 months working on the statue at the Wat Thammongkhol temple in Bangkok.

The boulder had originally been cut in Canada to reveal its inner quality and now comprised 14-ton and 10-ton pieces. The larger was turned into the now famous "Buddha in repose," while the other was carved into a statue of Quan Yin, the Chinese goddess of mercy. Pragmatic as well as ascetic, Abbott Phra Viriyang had the fragments of the jade blocks carved into 300,000 amulets. Imagine owning jewellery cut from the same boulder as the Great Lord Buddha! At $20 each, over $6 million was raised to cover expenses and to fund thousands of daycare centres throughout Thailand.

However, problems arose with the finish on the Buddha. Jade is a notoriously difficult material to polish because of its fibrous habit. In response to the Abbot's request, Canadians Kirk Makepeace and Tony Ritter spent two months over a period of two years, volunteering their time for free, to complete the polishing. As Kirk recalled, "When else would I have an opportunity to work on such a world-class project? And two months in Thailand is a nice change from the norm."[1]

In the early '90s, Tony Ritter (left) and Kirk Makepeace of Jade West worked on the surface finish of the world's largest jade Buddha at the Wat Thammongkhol temple in Bangkok. This massive sculpture was cut from a 32-ton jade boulder excavated near Dease Lake in northern British Columbia.

The reality was that sophisticated polishing of jade had evolved in Canada and that, combined with modern diamond technology, made it possible to complete the unique project. The abbot particularly liked the idea that Kirk was at the start and the finish of the jade project: it formed something of a cosmic cycle. The jade Buddha was completed in 1994 and installed along with the smaller sculpture of Quan Yin in a specially designed building within the temple grounds.

Appendix 1 : The Chemical Differences Between Jadeite and Nephrite

When chemistry advanced in the nineteenth century, it became apparent that the jades of Central America were not the same as the jades of China (and North American jades had not yet been discovered). Mineralogists classified Central American jade as jadeite, a mineral composed of sodium–aluminum and silica. Chinese jade turned out to be calcium–magnesium–iron and silicon and was given the name *nephrite* (from the Latin *nephriticus*, meaning kidney). Chemically they are quite distinct and mineralogically even less similar. Minerals are not just "puddings" of various elements but are structures built up in defined three-dimensional patterns. Both jade minerals are chains of silicon and oxygen atoms. Nephrite consists of double chains, while jadeite consists of single chains only.

Jadeite may be written as:

$$Na\ AL\ Si_2O_8$$

Jadeite from Burma is much prized in the Orient and is a superior lapidary material in its finest grades, but when compared to the recently discovered Polar Jade from B.C.'s Dease Lake area, there is little choice: Polar Jade wins out. Commercial deposits of jadeite are rare. They occur in numerous locations as a minor constituent of certain rocks but rarely in quantity or with the quality of the Burmese material. Mineralogically nephrite is composed mainly of tremolite–actinolite, a member of the amphibole family of minerals, characterized by long double chains of silica tetrahedrons linked by calcium, magnesium, and hydroxyl atoms/ions.

It may be depicted as follows:

$$Ca_2\ (Mg,\ Fe\ )_5\ (OH\ )_2\ Si_8O_{22}$$

The pure calcium mineral is tremolite, the pure manganese mineral is actinolite. In nature, however, you are not likely to find these compositions, but a blend of the two. Both are characterized by their cleavages, which are 56^0 for the amphiboles as opposed to 90^0 for the jadeite–pyroxenes. Tremolites and actinolites both contain water molecules, whereas jadeite is anhydrous, meaning it contains no water. Heat a fragment of nephrite in a closed test tube, and steam forms on the glass. Not so for jadeite.

The best green nephrite is polished on diamond wheels today and takes an almost mirrored finish. It has a hardness of 6.5 on the Mohs Scale, and the pure green variety is free of chromite spots or other impurities and discolourations. Jadeite is less tough than nephrite, but slightly harder (7.0 on the Mohs Scale). It has a fine granular structure that varies in hardness. This, in turn, means that polished surfaces have a slightly dimpled appearance upon close examination. However, the use of diamond polishing methods has reduced this effect, so it is not as readily apparent as it used to be.

Ultraviolet properties

Although both jade and emerald owe their colour to the presence of chromium, green jadeite does not show red under a Chelsea filter nor under long- or short-wave ultraviolet light. However, dyed jadeite (notably lavender) shows orange under long-wave UV light.[1]

Geological formation

As nephrite is a special textural form of the common mineral tremolite, the first question is how is tremolite produced, and secondly, how do we account for the special texture? Nephrite may be produced by two different processes. The less common method is the thermal metamorphism of dolomites (limestone) containing silica impurities, as a three-step process:

1. $2CaMg(CO_3)_2 + SiO_2 = 2CaCO_3 + Mg_2SiO_2 + 2CO_2$
   dolomite　　　quartz　　　calcite　　forsterite　carbon dioxide

2. The forsterite plus groundwater yields serpentine.
   $3Mg_2SiO_2 + 2H_2O = (3Mg0_2SiO_2.2H_2O)$
   forsterite　　water　　　serpentine

3. Tremolite then arises from the metasomatic* alteration of serpentine at the contact with siliceous rocks. This can be shown chemically as:
   $5(3Mg0_2SiO_2.2H_2O) + 14 SiO_2 + 6CaO = 3(2CaO).5MgO.8SiO_2.H2O + 7H_2O$
   　　serpentine　　　silica　　　　　　　　　tremolite　　　　　water

　*where minerals are changed chemically by the invasion of other　minerals (usually including water)

Under different conditions of temperature and pressure, different minerals may be stable along the serpentine contact and may produce deposits of idocrase, hydrogarnet, wollastonite, or talc. Commonly these occur as light-coloured or pale "whiterock" alterations and can be spotted from the air, making prospecting for possible nephrite deposits easy and rapid, provided, of course, you can see the ground. After a fire is the best time, as the ground cover is burned off and the white wollastonite and talc are particularly obvious.

However, nephrite is more than just a mixture of chemicals. To account for the special texture requires some unique conditions to prevail during the crystallization of the mineral fibres. These conditions require a stress-free environment and rapid crystal growth from innumerable centres. Further, the growth of the fibres must be entirely random. It has been postulated that the texture is simply inherited from the replacement of serpentine, which has this texture to begin with. If so, the explanation of the texture of the serpentine requires that crystallization also starts from innumerable centres in a static environment.

Appendix 2: World Jade Production

The average annual production from 1995 to 2000 in tons was 450: British Columbia, 200; Yukon, 20; Australia, 25; Russia (Siberia), 200; Taiwan, 0; U.S., 5; New Zealand, 0.

There is no doubt that the Siberian occurrences by themselves are enough to supply the world demand for jade, which might be in the order of a few hundred tons annually. There are also Australian reserves estimated at 50,000 tons. In Taiwan there are an estimated 80,000 tons, but much of this is not likely recoverable at reasonable cost. Canadian reserves may be 25,000 tons. In addition there are numerous small deposits in other parts of the world, which could add up to a few thousand tons.

In Bulletin V of the *Friends of Jade* an estimate of the world reserves of nephrite jade is recorded as a possible 90,000 inferred tons, distributed over nine countries. World consumption was at one time about 600[1] tons annually, but this figure has stabilized at about 300 tons at the time of writing.

The price per ton, or kilogram, depends on the size of the order and the average grade or quality of the shipment. Ton lots ship as a container load (20 tons) to Hong Kong, Taiwan, or Shanghai and may vary from $5,000 to $10,000 (U.S.). Smaller lots sold by the kilogram may vary from $5 to $50 (U.S.). Occasionally, very high-quality specimens may be more expensive, but the quantity sold will be small.

Mining jade

Although this is not a how-to book, here are a few tips for those interested in trying their luck at prospecting:

- Rodingite (which is white) in bedrock or boulders may indicate favourable geological conditions to find jade. Most nephrite deposits occur along or near the contact between ultramafic and metasedimentary rocks.
- Nephrite boulders have a rough, white/cream/brown/grey weathered surface, which makes them difficult to identify. The surface can appear grainy, like wood.
- After a fire has swept through an area, heat-exposed nephrite surfaces are often white and powdery.
- A hammer blow to nephrite usually doesn't leave a mark, and the hammer springs back with unexpected force because of jade's toughness.
- Large boulders need to be drilled or sawn to identify their worth.
- Once boulders have been found, move upslope, upstream, or up-ice to find the *in situ* deposit.

Jade mining in British Columbia progressed from cutting alluvial boulders and talus blocks with portable saws to unearthing jade lodes by excavator or bulldozer. The latter then involved stripping along the contact between serpentine and various country rocks in areas that yielded jade in surface exposures. The use of explosives is avoided as much as possible and generally is not necessary. No underground mining of jade has occurred in British Columbia.

Costs

Mining jade is a relatively inexpensive operation, and because of this, many small operators have been jade producers over the years. It is possible for one person with a helper to set up a portable diamond saw run by a gasoline motor, with a few hundred feet of hose to bring cooling water to the saw blade, to be in operation. Most operators soon find that sawing is a slow process, so two or three saws are usually operated at the same time.

Once a stockpile of sawn blocks amounts to a few tons, it's possible to have the production flown out by helicopter to the nearest road. In the Lillooet area, roads are always close at hand. In the central and northern districts, roads are farther from the mines. The actual cost of transportation is high but not excessive. There is then an added cost in getting the jade to one of the main highways, but this is fairly modest. Thereafter, road transport to Vancouver is reasonable.

One challenge that faced early jade miners was how to deal with very large boulders (greater than 100 tons). They cannot be sawn economically with a diamond saw, because very large saws (8 to 10 feet in diameter) are not available. One attempt to overcome this was made by New World Jade, who developed a wire saw on their Ogden Mountain property. This turned out to be a very slow and expensive process, so an alternative method was devised where holes were drilled and hydraulic splitters inserted. High pressure usually parted the rock, but jade's inherent toughness made that a slow process too.

Another method, deplored in North America but used extensively in Taiwan, was the use of explosives. This damaged the jade, however, and made it unattractive to the carving industry. At the Cassiar Asbestos Mine, explosives were used since they were a normal part of asbestos mining. Even when it was discovered that jade was a lucrative side product, the process was still continued. Most of the jade came out of the Cassiar pit in very large pieces, which did not pose a problem to the mine since it owned huge trucks, unlike most jade mining operations. These boulders were hauled to the mill site, where they were sold as rough and untested, without saw cuts or core holes. While this was convenient for the mine operators, it obviously meant that buyers had to take a chance, and the prices paid generally reflected their uncertainty. Nevertheless, the company sold over a million dollars' worth of jade.

Jade grades

Like all minerals, the outer skin or "rind" covering a jade boulder is extremely deceiving, even to the trained gemologist's eye. Jade is appraised by the appearance of sawn faces or drill cores through the rock. Colour is the greatest parameter; bright, "lively" green is the most highly prized, but there is a demand for very dark green to black material, and mottled colours may be acceptable in carving grades. Premium material must be free of fractures and inclusions of other minerals, although the presence of specks of chrome garnet, characteristic of Cassiar jade, is highly regarded by some buyers.

Four grades are commonly established:

Grade A
- Finest quality, good green colour, no fractures or inclusions. Material is homogeneous with no grain (structurally isotropic). Highly translucent and used in the gem trade.

Grade B
- Good colour in most of the material, a few fractures or inclusions, and little perceived graininess. Little material need be wasted. Used in the carving industry.

Grade C
- Inferior colour, numerous fractures or inclusions, not suitable for high-grade jewellery but may be used where variegation is acceptable. Usually structurally anisotropic (has a distinct grain in the rough). Used for low-end carvings like disks, slabs, bases, ashtrays, tiles, etc.

Grade D
- Non-commercial material, where the expense of salvage does not justify processing. Usually dumped on site.

Sellers set prices according to the grade, which may be subdivided into A+, B-, etc. In sales of large tonnage there is usually considerable scope for negotiation on both the grade and the unit price.

Future production

Nephrite jade is not a rare mineral, and many places in the world have the potential to become suppliers. There are two main geological conditions necessary for its formation. One is the contact between serpentine and other rocks capable of supplying calcium and silica. The other is the metamorphosis of dolomite under suitable conditions of temperature and pressure. Both of these conditions occur worldwide.

Jade is a depleting asset like all mineral commodities, but the relatively small demand and large reserves are likely to be maintained for some time to come, and the small but vital industry should continue into the distant future.

Glossary of Terms

accreted—grown together around a nucleus to form a whole.

actinolite—green fibrous amphibole mineral, found in crystalline schists and chemically similar to tremolite.

adit—horizontal shaft or passage used in mining to intersect a seam.

adze—tool similar to an axe, with an arched blade at right angles to the handle.

aggregate—mass of rock particles or mineral grains.

alluvial—deposited or formed by running water.

alluvium—deposits left by rivers or other moving water.

altiplano—high plateau in Bolivia and Peru, averaging 13,000 feet elevation.

amazonite—green variety of microcline feldspar, somewhat similar to jade in appearance.

amphibole—common rock-forming minerals found in igneous and metamorphic rocks.

andesite—fine-grained, dark volcanic rock.

anisotropic—having non-uniform properties: inhomogeneous.

anorthosite—plutonic igneous rock comprised mostly of plagioclase feldspar.

asbestos—term applied to a number of fibrous minerals, sometimes called "rock wool."

aventurine—semi-precious green gem of quartz or feldspar, superficially similar to jade.

bars—a unit of pressure measurement.

basalt—dark, fine-grained volcanic rock, very common in earth's crust.

batholith—extensive plutonic mass of deep earth origin.

bedrock—solid rock underlying loose material (gravel, soil).

Beringia—region of western Alaska, Bering Strait, and eastern Russia that remained uncovered during the last ice age.

beryl—important gemstone: green variety is emerald, blue is aquamarine, pink is morganite.

biotite—mineral of the mica group.

botryoidal—having the form of a bunch of grapes.

bowenite—yellow/green variety of serpentine, superficially resembling jade.

BP—before the present, measured in years.

cabbing—making cabochons on a grinding wheel.

cabochon—oval-shaped polished stone of varying size, used widely in semi-precious gemstone settings.

calcsilicate—minerals rich in calcium–magnesium, often containing epidote, wollastonite, vesuvianite, etc.

calcite—stable form of calcium carbonate and main constituent of limestone and marble.

californite—variety of vesuvianite, known as "California jade."

Canadian Shield—largest Precambrian rock region in the world.

Carboniferous—geological time between 360–290 MYA, further divided into Upper and Lower, or Pennsylvanian and Mississippian.

celt—stone chisel.

chalcopyrite—brass coloured copper–iron–sulfide, important source of copper ore.

char—northern sea trout.

chert—or flint: a dense, hard, silica-rich mineral used widely by Neolithic cultures.

chromite—one of the spinel group of minerals, found in mafic and ultramafic rocks.

chrysoprase—semi-precious, translucent green gemstone formed by micro-crystalline quartz.

chrysotile—fibrous white/grey/green mineral of the serpentine group.

cirque—natural amphitheatre formed in a hill or mountain.

cobble—rock, usually water-rounded, larger than a pebble, smaller than a boulder.

concretion—lump/nodule found in limestone, shale or sandstone.

contact zone (aureole)—region of mineral/chemical change where igneous processes transform the host rock.

Cordillera—western mountain spine of North America, including Coast and Rocky mountains.

Devonian—geological time between 410–360 MYA.

dolomite—an altered limestone where some of the calcium is replaced by magnesium.

drill core—solid rod of minerals removed when a hollow rock drill is brought up to the surface.

dunite—mineral comprised mostly of olivine, but may contain pyroxene, plagioclase, and chromite.

dyke—intrusive igneous rock that cuts across existing strata.

esker—long, narrow ridge of loose material formed by glacial meltwater.

feldspar—important group of rock-forming minerals, making up 60 percent of the earth's crust.

ferrules—metal strengthening ring used on joint or pipe.

flint—see chert.

float—loose rock found downslope from a mineral source.

footwall—mass of rock below a fault plane, ore body, or mineworking.

forsterite—white/yellow magnesium–silicate, belonging to olivine family.

fusible—the property of being able to go from solid to liquid when heat is applied (melt).

gabbro—intrusive rock formed mostly of plagioclase and pyroxene.

galena—dense metallic lead–sulfide mineral, primary ore of lead.

gneiss—foliated metamorphic rock, often coarse-grained, commonly with biotite and hornblende.

graphite—dark to black form of natural carbon, often found in metamorphic rocks.

greenstone—any green, weakly metamorphosed igneous rock.

greywacke—very hard, dark grey/green coarse-grained sandstone.

grossular—white/yellow member of the garnet group, often found at contact zones in limestone.

habit—general appearance of a crystal: cubic, prismatic, etc.

hanging wall—overlying side of a fault, ore body, or mineworking.

hornblende—a silicate mineral, most common component in amphibole group.

idocrase—see vesuvianite.

igneous—solidified rock formed from magma generated deep within the earth.

in situ—positioned in the place where it was formed.

intrusive—igneous rock formed from magma that cools below the earth's surface.

isotropic—equal properties in all directions, uniformly formed.

jadeite—monoclinic mineral of the pyroxene group.

Jurassic—geological period between 210–140 MYA.

kimberlite—ultramafic igneous rock that forms deep within the earth, often reaching the surface in vertical pipes.

labradorite—plagioclase feldspar occurring in base igneous or high grade metamorphic rocks.

lapidary—relating to the craft of cutting and polishing stone.

lapis lazuli—rich blue, precious gemstone formed from lazurite, calcite, and sometimes pyrite.

laterite—soil residue formed from weathered iron–oxides.

lazulite—a magnesium–iron–aluminum phosphate, found in high-temperature metamorphic rock.

lazurite—blue mineral of the sodalite group and main constituent of lapis lazuli.

lens—ore body that is thick in the centre and thin at the ends.

limestone—sedimentary rock comprised mostly of calcite (calcium carbonate).

listwanite—mineral comprised mostly of quartz–carbonate.

lode—a mineral deposit.

lode claim—mineral claim entitling holder to subsurface minerals on the property.

mafic—usually dark magnesium- and iron-rich minerals like olivine, pyroxene, amphibole.

magma—molten rock generated deep within the earth.

magnesite—magnesium carbonate, often formed when the calcium in calcite is replaced by the magnesium ion, harder and denser than calcite.

magnetite—strongly magnetic, shiny black iron–oxide, an important ore of iron.

malachite—colourful green, semi-precious copper–carbonate gemstone.

marble—hard crystalline rock formed from metamorphosed calcite and/or dolomite.

Mesoamerica—region of Central America covering Mexico to Panama.

metachert—metamorphosed chert, where original chert is still recognizable in the material.

metamorphic—important class of rock, whose properties have changed over time by heat, pressure, etc.

metasedimentary—metamorphosed sedimentary rocks still recognizable as sedimentary in origin.

metasomatism—metamorphic process whereby a mineral is changed by chemical replacement.

mica—group of minerals common in igneous and metamorphic rocks, generally plate-like in form.

Mississippian—term for Lower Carboniferous period (used within North America) 360–350 MYA.

Mohs—mineralogist who proposed a method (in 1922) of measuring rock hardness on a scale from 1 to 10.

mookite—dense, colourful semi-precious silica gemstone found in Western Australia.

moraine—loose rock material carried and deposited by glacier.

motherlode—original source of minerals or ore body.

muezzin—Muslim crier who proclaims the hours of prayer.

MYA—geological shorthand for "million years ago."

Neoloithic—late Stone Age period, when ground/polished stone tools predominated.

nephrite—extremely tough, compact tremolite/actinolite amphibole.

nodule—rounded/lumpy concretion.

olivine—yellow/green rock-forming ultramafic mineral common in basalt and gabbro.

overburden—sediments overlying a stratum of interest.

Pennsylvanian—term for Upper Carboniferous period (used within North America) 320–290 MYA.

Paleolithic—early Stone Age period when stone tools were formed by chipping.

pelite—fine-grained mudstone.

Permian—geological time between 290–250 MYA.

phyllite—silver/grey metamorphic rock, texture between slate and schist.

placer—surface deposit containing minerals of economic value.

placer claim—mineral claim entitling holder to surface minerals on the property.

plagioclase—important group of feldspars (albite–anorthite).

Pleistocene—geological time period of Quaternary Period between 1.6–0.01 MYA.

plug and feather—non-explosive mining technique using drill holes and expanding wedges.

plutonic—rocks formed at depth in the earth.

porphyry—rock containing large crystals within a host mass of fine-grained material.

Precambrian—geological time period from earth's formation until about 570 MYA.

provenance—place of origin of rocks or minerals.

pyroxene—common rock-forming group of minerals found in basic igneous and high-grade metamorphic rocks.

quartz—one of the commonest rock-forming minerals, comprising silicon–oxide.

radiocarbon dating—technique to determine age of artifact from 60000 BP to present.

reaction zone—region in which two or more minerals contact and change.

rhodonite—strikingly pink, semi-precious, manganese–silicate gemstone.

rockhound—amateur prospector or lapidary hobbyist.

rodingite—"white rock" calesilicate formed by alteration of gabbro in contact with serpentine.

sandstone—sedimentary rock formed from sand, silt, or clay grains.

schist—metamorphic rock exhibiting grain-like or banded patterning.

schistose—schist-like.

sedimentary—important rock group formed by consolidation of waterborne sediments.

selenite—colourless form of crystalline gypsum (calcium–sulfate).

serpentine—common group of rock-forming minerals, often fibrous, derived from low-grade metamorphism.

serpentinite—metamorphic rock formed almost entirely from serpentine minerals.

Silurian—geological time between 439–408 MYA.

shale—fine-grained sedimentary rock formed from silt, clay, or sand.

slate—fine-grained mudstone produced by low-grade metamorphism, showing pronounced cleavage.

smithsonite—ore of zinc, often strongly coloured, often found in sulfide deposits.

soapstone—soft metamorphic rock comprising mostly talc, easily carved using steel tools.

sodalite—blue semi-precious gemstone belonging to the feldspar group.

spall—remove by splintering or chipping.

spinel—a group of magnesium-rich hard crystalline minerals of various colours.

spodumene—precious gemstone when found in crystalline form, belonging to the pyroxene group.

stratum—layer of sedimentary rock (plural: strata).

stromatolite—calcium-rich formations left by bacteria, some of great age (3,000 MYA).

talc—very soft whitish mineral formed by alteration of magnesium silicate or ultramafic rocks.

talcose—talc-like.

talus—coarse debris caused by weathering of cliff face above.

tectonic plate—large segment of the earth's crust that moves as a continuous mass.

terrane—region or zone of rocks that may be regarded as being similar to each other.

tickle—Eastern Canadian colloquialism for sea channel.

tiger-eye—semi-precious silica gemstone metamorphosed from fibrous asbestos minerals.

tiger-iron—semi-precious gemstone comprising bands of tiger-eye, jasper, and hematite.

tremolite—white/green mineral of the amphibole group, source of asbestos-related materials.

Triassic—geological time between 250–210 MYA.

ultramafic—containing more than 90 percent mafic materials.

vesuvianite—green/brown, semi-precious gemstone formed through metamorphism of impure limestones.

volcanics—fine-grained igneous rocks formed from magma and cooled rapidly at earth's surface.

vulcanism—process associated with magma and/or steam being ejected by a volcano.

wollastonite—pale calcium–silicate mineral typically formed at contact with calcium-rich rocks.

Endnotes

Preface

1. S.F. Leaming, *Jade in Canada*. This publication laid the scientific groundwork for the burgeoning jade industry in Canada and, more specifically, British Columbia.

The Canadian Jade Story

1. Holland, "Jade in British Columbia." In 1961 this report laid out the prospecting requirements that enabled miners to find nephrite in the province more easily.
2. Bud Davidson, (son of Bill Yarmack), personal communication, May 2004.
3. Overend, "Jade by the ton."
4. Street, "New jade (*in situ*) deposit found in British Columbia."
5. Moira Farrow, "That cigarette did it," *Vancouver Sun*, November 27, 1971, p. 25.
6. "Jade jackpot for B.C. family," *The Province*, September 13, 1969, p. 9.
7. "Six-Ton Jade Going Into Pool," *Vancouver Sun*, March 16, 1967, p. 13.
8. Moira Farrow, "That cigarette did it."
9. "Jade find could be rich deposit," *Surrey Columbian*, February 21, 1970.
10. "Structural Materials & Industrial Minerals," *B.C. Minister of Mines and Petroleum Resources Annual Report*, p. 129. Victoria, B.C.: Queen's Printer, 1960.
11. *Surrey Columbian*, October 23, 1968.
12. Win Robertson, personal communication, June 27, 1996.
13. Ibid.
14. Fraser, M., "The jade mines of B.C."
15. Win Robertson, personal communication, June 27. 1996.
16. "Exotic bids made for B.C. jade," *The Province*, October 8, 1969, p. 21.
17. Fraser, M., "The jade mines of B.C."
18. Fish, "East and West meet at B.C. jade mine."
19. Bev Christensen, "How I found tons of jade," *Vancouver Sun*, April 7, 1971, p. 30.
20. Win Robertson, personal communication, 1996. Win wrote extensively on her experiences in the jade industry.
21. Fraser, J.R., "Nephrite in British Columbia."
22. L. Warren, personal communication, 1996.
23. Jane Becker, "Green boulder started it all," *Vancouver Sun*, January 24, 1970, p. 30.
24. "Dempsey Mines jade start up," *The Province*, February 22, 1971.

25. Gerry Davis, personal communication, March 1985.
26. Fraser, M., "Nephrite Jade, The Stone of Heaven."
27. Ben Seywerd, personal communication, 1982.
28. Kirk Makepeace, personal communication, September 2003. Kirk is chief executive officer of the Jade West Group and a key player in the industry. He has contributed many stories of the early days of prospecting.
29. Ward, "Jade: Stone of Heaven." This feature article in *National Geographic* generated a lot of interest in what, until then, had been regarded as a "foreign" gemstone.
30. Ibid.
31. Dawson, "Report on the exploration of the Yukon District, NWT and adjacent northern portion of British Columbia 1887." This was the first serious geological assessment of the region. Dr. Dawson covered much ground in a short time, and missed very few of the major mineralogical features.
32. "Jade miners report big new find," *The Province*, July 21, 1971, p. 13.
33. Porter, "Skin diving for jade."

Jade Craft Arrives in North America

1. Kunz, "Gems and precious stones of Mexico."
2. Emmons, *Notes and Monographs.*
3. Smith, *American Museum of Natural History Annual Report.*
4. Dawson, "Report on the exploration of the Yukon District."

The Science of Jade

1. Hughes and Kouznetsov, "From Russia with jade." This article recounts a major new jadeite discovery in northern Russia. Given the poor infrastructure and local corruption, it will be some time before this source is brought onstream.
2. Bradt, et al., "The toughness of jade" and Belyk, et al., "The mechanical properties of B.C. jade."

Jade in North America and Europe

1. Emmons, *Notes and Monographs.*
2. Nelson, ed., "On the source of the jade implements of the Alaskan Inuits."
3. George T. Emmons, *Notes and Monographs*, No. 35, Museum of the American Indian, (New York: The Heye Foundation, 1923).
4. Ibid.
5. Anderson, "Asbestos and jade occurrences in the Kobuk River region, Alaska."
6. Proctor, "Jade beyond the Arctic Circle" and Munz, "Hundreds of tons of jade."
7. Proctor, "Jade beyond the Arctic Circle."
8. Smith, *American Museum of Natural History Annual Report.*
9. Ream, "Nephrite in Washington."
10. Coleman, "Low temperature reaction zones and alpine ultramafic rocks of California, Oregon and Washington."
11. Shepherd, "Wyoming jade."
12. Ruff, "The Jade Story–Part 7." Part of a major review article on nephrite and jadeite that was serialized in the *Lapidary Journal.*
13. Porter, "Skin diving for jade."

14. Ruff, "The Jade Story–Part 7."
15. Ibid.
16. Reed, "Guatemala's Olmec jade."
17. Ruff, "The Jade Story–European."
18. Fischer, "On stone implements in Asia."
19. Kunz, *Investigations and Studies in Jade*.
20. Geiss, "Jade in Switzerland."
21. Tucan, "Jadeites from Aljagica in South Serbia."
22. Gievlenko, et al., *Geology of gemstone deposits*.

Hunting the Stone of Heaven: A Personal Odyssey

1. Yule, *Cathay and the way thither*.
2. Carpenter, "Taiwan nephrite jade."
3. Nieder, "Nephrite jade in the Republic of Korea."
4. Barnes, et al., "A World Review of Nephrite Jade—Geology, Production and Reserves." *Friends of Jade*, Vol. 5, 1988.
5. Chalmers, "New occurrences of gem minerals in Australia."
6. Finlayson, "The metallogeny of the British Isles."
7. Beck,"A new development in understanding the prehistoric usage of nephrite in New Zealand." This paper dealt with the Maori technique of fire-treating semi-nephrite to harden and colour it.

Jade in Canada's Courts

1. "Jade trial takes two days ..." *Cassiar Courier*, June 1982, p. 1.
2. "30 tons of stolen jade nets two years in jail for Cassiar-area pair," *Vancouver Sun*, July 29, 1982.
3. "$8,000 jade vanishes," *Vancouver Sun*, December 20, 1972, p.29.
4. "Jade theft loss put at $40,000," *The Province*, March 7, 1975, p. 21.
5. "Jade load suspect," *Vancouver Sun*, September 8, 1970.
6. Bob McMurray, "$5 million jackpot on jade killed by government," *The Province*, October 23, 1979.
7. "Chip off the old block," *Kelowna Courier*, March 12, 1982.
8. "Jade boulder stolen," *Vancouver Sun*, March 10, 1982, p. A2.
9. "Prospectors clash," *The Province*, Sept 22, 1980.
10. "Art Centre proposed," *Vancouver Sun*, April 24, 1975.

The Jade Sculptors

1. Kirk Makepeace, May 2004. Kirk provided written recollections of Dease Lake mining activities.

Appendix 1

1. Koivula, "Some observations on the treatment of lavender jadeite."

Bibliography

Anderson, Eskil. "Asbestos and jade occurrences in the Kobuk River region, Alaska." Pamphlet No.3, Alaska Department of Mines, 1945.

Barnes, L.C., et al. "A World Review of Nephrite Jade—Geology, Production and Reserves." *Bulletin of the Friends of Jade*, Vol. 5, 1988.

Beck, Russell J. *New Zealand Jade: The Story of Greenstone.* Wellington, NZ: A.H. & A.W. Reed Ltd., 1970.

———. *New Zealand Jade.* Wellington, NZ: A.H. & A.W. Reed Ltd., 1984.

———. "A new development in understanding the prehistoric usage of nephrite in New Zealand." *Archaeological Studies of Pacific Stone Research.* F. Leach & J. Davidson, eds., British Archaeological Reports, International Series 104. Oxford, UK: ArcheoPress, 1981.

Beck, Russell J. with Maika Mason. *Mana Pounamu New Zealand Jade.* Auckland, NZ: A.H. & A.W. Reed Ltd., 1984.

Belyk, G.E., et al. "The mechanical properties of B.C. jade." Undergraduate report. University of British Columbia, Faculty of Engineering, February 29, 1973.

Bradt, R.C., et al. "The toughness of jade." *American Mineralogist*, Vol. 58, pp. 727–732, 1973.

Carpenter, E.K. "Taiwan nephrite jade." *Lapidary Journal*, p. 942, October 1968.

Chalmers, R.G. "New occurrences of gem minerals in Australia," *Journal of Gemmology*, Vol. 12, No. 7, July 1971.

Coleman, R.G. "Low temperature reaction zones and alpine ultramafic rocks of California, Oregon and Washington." U.S. Geological Survey Bulletin, p. 1,247, 1967.

Dawson, George M. "Report on the exploration of the Yukon District, NWT and adjacent northern portion of British Columbia 1887." Ottawa, ON: Geological Survey of Canada, 1888.

Day, Elva. "Jade hunting in New Zealand." *Lapidary Journal*, p. 726, August 1983.

Emmons, Lt. George T. *Notes and Monographs*, No. 35. Museum of the American Indian (Heye Foundation), 1923.

Finlayson, A.M. "The metallogeny of the British Isles." *Quarterly Journal of the Geological Society of London*, p. 281, 1909.

Fish, Richard H. "East and West meet at B.C. jade mine." *Northern Miner*, p. 44, November 29, 1973.

Fischer, L.H. "On stone implements in Asia." Private paper, 1884.

Fraser, J.R. "Nephrite in British Columbia." Ph.D. thesis, Department of Geology, University of British Columbia, 1972.

Fraser, Marilyn. "Nephrite Jade, The Stone of Heaven." *Canadian Rockhound*, Vol. 4, No. 2, 2000.

———."The Jade Mines of B.C." *Canadian Rockhound*, Vol. 4, No. 2, 2000.

Geiss, H. "Jade in Switzerland." *Bulletin of the Friends of Jade*, No. 8, Summer 1994.

Gievlenko, E.I., et al. *Geology of gemstone deposits*. Moscow: Science Publishing House, 1976.

Holland, S.S. "Jade in British Columbia." B.C. Minister of Mines and Petroleum Resources Annual Report, pp. 119-126, 1961.

Hughes, R., and Nickolai Kouznetsov. "From Russia with Jade." *GemKey Magazine*, Vol. 3, No. 1, pp. 58–66, 2000.

Keverne, Roger, ed. *JADE*. London, UK: Anness; New York, NY: Van Nostrand Reinhold, 1996.

Koivula, J.I. "Some observations on the treatment of lavender jadeite." *Gems & Gemology*, 28 (1), p. 32, 1982.

Kunz, Dr. George F. *Investigations and Studies in Jade*. New York, NY: DeVinne Press, 1906.

———."Gems and precious stones of Mexico." *Transactions of the American Institute of Mining Engineering*, 1901.

Leaming, S.F. *Jade in Canada*. Geological Survey of Canada Paper 78/19, 1978.

———."Jade fields of Australia." *Cab & Crystal*, p. 2,124, Winter 1991.

Lee, Gwen, and Don. *Rivers of Gold: A True Yukon Story*. Vancouver, BC: Peanut Butter Publishing, 1999.

Munz, William. "Hundreds of tons of jade." *Lapidary Journal*, p. 18. April 1970.

Nagle, C. "Flaked stone procurement and distribution in Dorset Culture sites along the Labrador Coast." *Publications in Archaeology*, St. John's, NF: Memorial University, April 1984.

Nelson, E.W., ed. "On the source of the jade implements of the Alaskan Inuits." *The Conference Proceedings of the U.S. National Museum*, Vol. VI, pp. 426-427, 1883.

Nieder, Captain A.E. "Nephrite jade in the Republic of Korea." *Lapidary Journal*, pp. 2,374–2,380, March 1982.

Overend, Miles. "Jade by the ton." *Canadian Business*, pp. 62–69, October 1968.

Owen, Larry. "The treasure of Ogden Mountain." *Lapidary Journal*, p. 114, April 1972.

Owen, Margaret. *So We Bought the Town*. Vancouver, BC: Mitchell Press, 1977.

Porter, Alyce. "Skin diving for jade." *Lapidary Journal*, August 1958.

Proctor, William. "Jade beyond the Arctic Circle." *Lapidary Journal*, p. 598, May 1977

Purvis, Ron. *T'shama*. Surrey, B.C.: Heritage House Publishing Co., 1994.

Ream, Larry. "Nephrite in Washington." *Lapidary Journal*, Vol. 29, No. 9, p. 1,748, December 1975.

Reed, Christina. "Guatemala's Olmec jade." *Geotimes*, American Geological Institute, August 2002.

Richardson, Sir John. *Arctic Searching Expedition*. London, UK: Longman, Brown, 1851.

Ruff, Elsie. "The Jade Story–Part 7." *Lapidary Journal*, Vol. 15, No. 4, p. 414, October 1961.

———."The Jade Story–European." *The Journal of Gemmology*, Proceeding of the Gemmology Association of Great Britain, Vol. 4, No. 1, January 1953.

Shepherd, Kenneth R. "Wyoming jade." *Lapidary Journal*, p. 1596, March 1972.

Smith, H.I. *American Museum of Natural History Annual Report*, Washington, DC: 1900–01.

Street, Harry J. "New jade (in situ) deposit found in British Columbia." *Lapidary Journal*, p. 246, April 1966.

Tucan, Fran. "Jadeites from Aljagica in South Serbia." *Neues Jahrbuch für Mineralogie und Geochemie*, Vol. II, pp. 764–765, 1984.

Ward, Fred. "Jade: Stone of Heaven." *National Geographic Magazine*, Vol. 172, No. 3, September 1987.

Ward, Fred and Charlotte. *Jade*. Bethesda, MA: Gem Book Publishers, November 1996.

Yule, Sir Henry. *Cathay and the way thither*. Translated 1866; first published London, UK: Hakluyt Society, 1915.

Newspapers

The Province
Northern Miner
The Vancouver Sun
Cassiar Courier
Kelowna Courier
Surrey Columbian

Journals, periodicals, and magazines

American Mineralogist
American Museum of Natural History Annual Report
BookNews
B.C. Minister of Mines and Petroleum Resources Annual Report
Bulletin of the Friends of Jade
Cab & Crystal
Canadian Business
Canadian Rockhound
GemKey Magazine
Gems & Gemology
Geotimes
Lapidary Journal
National Geographic Magazine
Neues Jahrbuch für Mineralogie und Geochemie
Northern Miner
Publications in Archaeology
Quarterly Journal of the Geological Society of London
The Conference Proceedings of the U.S. National Museum
The Journal of Gemmology

Index

Photo Credits

STANLEY FRASER LEAMING worked across Canada as a field geologist for over a decade before joining the Geological Survey of Canada in 1960. He has travelled the world researching jade, its geology, and its uses. His published works include *Rock and Mineral Collecting in British Columbia*, *Jade in Canada*, and the *Guide to Rocks & Minerals of the Northwest* (with his son Chris). Stan lives in Summerland, B.C.

RICK HUDSON is part-owner of Mineral World, a geological attraction in Sidney, B.C., that draws over 100,000 visitors annually, and he teaches earth sciences in schools. He was the editor of the *Western Canadian Gemstone* newsletter for five years and is a contributing writer to the on-line magazine *Canadian Rockhound*. He is also the author of the two-volume book *Field Guide to Gold, Gemstone & Mineral sites of British Columbia*. Rick lives near Victoria, B.C.